Macmillan Encyclopedia of the Environment

Macmillan Encyclopedia of the
ENVIRONMENT

VOLUME 3

General Editor
Stephen R. Kellert

Associate Editors
Matthew Black

Richard Haley

Macmillan Library Reference USA
New York

Developed, Designed, and Produced by Book Builders Incorporated

Macmillan Library Reference
1633 Broadway, New York, NY 10019-6785

Library of Congress Catalog Card Number: 96-29045

Printed in the United States of America

Library of Congress Cataloging-in-Publication Data

Macmillan encyclopedia of the environment.
      p.   cm.
    "General editor, Stephen R. Kellert"—P. iii.
    Includes bibliographical references and index.
    Summary: Provides basic information about such topics as minerals, energy resources, pollution, soils and erosion, wildlife and extinction, agriculture, the ocean, wilderness, hazardous wastes, population, environmental laws, ecology, and evolution.
    ISBN 0-02-897381-X (set)
    1. Environmental sciences—Dictionaries, Juvenile.
[1. Environmental protection—Dictionaries. 2. Ecology—Dictionaries.]
    I. Kellert, Stephen R.   96-29045
    GE10.M33 1997   CIP
    333.7—dc20   AC

Photo credits are gratefully acknowledged in a special listing in Volume 6, page 102.

G

Gaia Hypothesis

▶The belief of James E. LOVELOCK that "the earth's living matter, air, OCEANS, and land surface form a complex system which can be seen as a single organism and which has the capacity to keep our planet a fit place for life." The "Gaia" concept (named after the ancient Greek goddess of Earth) was introduced in 1974 by James E. Lovelock and Lynn Margulis, in the article "Biological Modulation of the Earth's Atmosphere."

The idea was developed further by Lovelock in two books, *Gaia: A New Look at Life on Earth* (1979) and *The Ages of Gaia* (1988). Lovelock, like other scientists before him, noted that organisms have greatly affected the makeup of Earth's ATMOSPHERE. Living things also affect other aspects of the planet, such as temperature and the movement of nutrients. These and other observations led Lovelock to propose that organisms can *regulate* conditions in the air, land, and water in a way that promotes the overall well-being of life on Earth. He compared this ability to the internal regulation, or *homeostasis*, that maintains the health of an organism and proposed that Earth itself was a living organism.

The Gaia hypothesis has provoked a great deal of criticism. It has been argued that interactions between life and the ENVIRONMENT on Earth are not at all precisely regulated or generally beneficial to the biota, or flora and fauna, of a region; that is, the planet does not really maintain its own health the way a living organism does. Also, as even Lovelock points out, the hypothesis is not testable. There is no other Earth we can use for experiments to prove the idea right or wrong. Thus, the hypothesis cannot be evaluated by the usual scientific approach.

Whether it is generally accepted or not, the hypothesis provokes questions about how "life" and "organisms" are defined. It may also stimulate more research on large-scale interactions between living communities and their physical surroundings. [*See also* BIOLOGICAL COMMUNITY; BIOSPHERE; and ECOSYSTEM.]

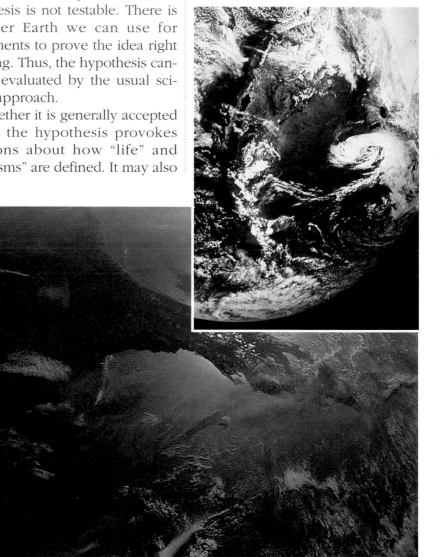

◆ Is Earth an organism? An organism is capable of homeostasis, the process of maintaining a healthy condition. Can Earth's species, the air, land, and water act as one organism and maintain a healthy condition?

Galápagos Islands

▌A group of volcanic islands in the Pacific Ocean. The Galápagos Islands lie on or near the equator approximately 600 miles (960 kilometers) off the coast of Ecuador, the country to which they belong.

The Galápagos Islands are officially known as the *Archipiélago de Colón*. The Galápagos Island group consists of 5 large islands, 8 small islands, 42 islets, and 26 rocks, which cover an area of 3,029 square miles (7,874 square kilometers). It was first sighted in 1535 by Thomás de Berlanga, a Spanish navigator en route to Panama. The islands were named for the giant tortoises living there—*galápagos* is Spanish for "turtle".

The Galápagos **archipelago** is crossed by five different OCEAN CURRENTS. These currents control the

◆ The Galápagos marine iguana is the only known marine lizard species. It feeds on seaweed.

CLIMATE and have played an important part in the development of the unusual life forms for which the Galápagos Islands are famous. Although you would expect hot, tropical WEATHER so close to the equator, a cool ocean current moving north along the South American coast and currents flowing south from North America's western coast provide cool temperatures to the region. Over time, such currents carried a diversity of WILDLIFE from various parts of the world to the Galápagos Islands.

ANIMAL AND PLANT LIFE

At sea level, the Galápagos Islands are barren and rugged. However, each island has its own characteristic SPECIES of living things, especially BIRDS, lizards, and tortoises. On fields of volcanic lava, animals that blend in with their surroundings hide. Multicolored marine iguanas blend in with rock that is covered with LICHENS. Sea lions seem to blend in with the brown and black volcanic rubble. Island mountains

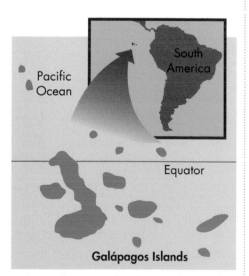

◆ The Galápagos Islands sit in the Pacific Ocean on or near the equator, some 600 miles (960 kilometers) off the coast of Ecuador.

that rise to high elevations are covered by a rich, subtropical FOREST.

The Galápagos Islands have been called the "Enchanted Islands" by people who marveled at the archipelago's animals, many of which are found nowhere else on Earth. Such animals include huge tortoises, many weighing more than 500 pounds (225 kilograms), rare flightless **cormorants**, and 4-foot-long (1.2-meter-long) marine iguanas that eat seaweed. The marine iguana is unique as the only known marine lizard species.

Other Galápagos inhabitants include scarlet crabs that evolved from a species that lived in the Atlantic Ocean. The crabs separated from their crustacean relatives about 35 million years ago. The

◆ Huge Galápagos tortoises may weigh more than 500 pounds. Once taken aboard sailing ships and used for food, they are today an endangered species.

many unique species convinced some people that these islands, which are also home to millions of herons, frigate birds, penguins, and sea birds known as boobies, were a land lost in time.

During the nineteenth century, whalers often captured giant Galápagos turtles to take on long journeys as ready supplies of fresh meat for their crews. They kept the turtles in the ship's hold, where the animals could survive for up to a year without food. After Ecuador claimed the islands in 1831, the unique wildlife came under the country's protection. In 1959, Ecuador declared the Galápagos Islands a NATIONAL PARK. In 1979, the United Nations established the archipelago as a World Heritage Site.

LIVING THINGS— CREATED OR DEVELOPED?

In 1835, British naturalist Charles Robert DARWIN visited the Galápagos Islands during the voyage of the H.M.S. *Beagle*. While there, he studied the islands' wildlife. His scientific observations—such as the discovery that ground finches on one island had strong beaks and ate large seeds, whereas those of a very similar species living on another island had smaller, weaker beaks and ate INSECTS—led him to the study of animal ADAPTATION.

When Darwin returned to England, he began evaluating his findings to determine why birds of the same species differed from island to island. He noted that the islands of the Galápagos were just far enough apart to prevent birds of the same species from flying from one island to the other. From this and other observations, Darwin developed his theory of EVOLUTION.

THE GALÁPAGOS ISLANDS TODAY

Because of the rugged, volcanic landscape of the Galápagos Islands, only about 6,200 people live there. Scientific study of the wildlife continues. Each year thousands of tourists come to see the unique PLANTS and animal life in the protected wilderness area. To maintain and protect the Galápagos Islands, ECOTOURISM has been strictly regulated. Concern about the human impact on the ENVIRONMENT has led to increased monitoring of heavy use areas and GARBAGE and other waste products that present potential threats to the environment. Some of the major environmental problems in the Galápagos today include the destructive impact of certain herbivore species that were introduced to the islands, overfishing, and POACHING. [*See also* BIODIVERSITY; BIOLOGICAL COMMUNITY; ECOSYSTEM; ENVIRONMENTAL ETHICS; and SEALS AND SEA LIONS.]

Garbage

▌Disposable waste, usually from homes and businesses, that is BIODEGRADABLE. Besides food scraps, the average household may produce burnable SOLID WASTE like paper, cardboard, leaves, and rags as well as nonburnable waste such as glass and plastic bottles, plastic containers, and aluminum or tin cans. Some people also throw out old clothing, kitchen utensils, furniture, and books as garbage.

Most garbage will eventually rot away, erode, or disintegrate, but the length of time it takes to do so differs depending on the type of garbage. Conditions such as how large an object is, what it is made of, and the temperature and moisture around it help determine how long it takes for the material to break down.

THE LANGUAGE OF THE ENVIRONMENT

garbologist a scientist who studies garbage.

mummified preserved by drying out, withering, or shriveling up.

◆ Garbage is collected by sanitation workers and brought to a transfer station before it is eventually brought to sanitary landfills.

GARBAGE PROBLEM

The average American throws out more than 3.3 pounds (1.5 kilograms) of stuff per day—enough in one year to produce a pile as tall as a 13-story building! In the past, people gave little thought to what happened to garbage after sanitation workers picked it up from the curbside. Usually, the garbage was hauled to an open dump and unloaded. Piles of smelly, rotting materials attracted disease-carrying flies and rats, so beginning in the 1920s, people began to bury garbage in sanitary LANDFILLS.

SANITARY LANDFILLS

The concept behind sanitary landfills is scientifically based. Buried material will rapidly decompose, making a richer SOIL for future use. However, decay requires adequate amounts of OXYGEN and moisture to encourage the growth of the BACTERIA necessary to break down the materials. Often, in sanitary landfills air and water could not get through the piled-up garbage.

When **garbologist** William Rathje, head of the University of Arizona's Garbage Project, dug up LANDFILLS in the 1980s, he found that chemical changes were slower than expected. Rathje discovered newspapers buried more than 50 years that were yellowed but as easy to read as the day they were printed. Foods like corn on the cob were perfectly recognizable, though not appetizing, and some hot dogs had actually been **mummified**. Glass and plastic objects, the garbologist noted, were relatively unchanged over time.

Other problems arose from landfills built in the first half of the twentieth century. They had been built on sites no one used or wanted to develop. They were built usually on low land because it was easier to fill a hole than build a hill. However, many such sites were ecologically fragile marshlands, where countless BIRDS and other animals lived. Not only did the WILDLIFE lose its HABITAT, but nearby land and water became polluted.

When a landfill rose above sea level, it was closed. Builders then came to construct new neighborhoods on the supposedly enriched soil. Instead, the land and water had become polluted with unsafe, sometimes toxic, materials. For example, heavy metals from batteries buried in landfills seeped through the soil to pollute ground-

◆ Americans generate a lot of garbage—on the average more than 3.3 pounds (1.5 kilograms) per person per day.

water. Or leaking chemicals from illegally stored containers showed up in yards and homes, threatening the lives of residents, as in the LOVE CANAL landfill area of Niagara Falls, New York.

Today, a landfill is planned to have the smallest possible effect on the ENVIRONMENT. But Earth is running out of space on which to build landfills. With limited space for landfills, interstate transport of garbage to communities with enough space to bury garbage has become a problem. Garbage that is hauled by truck or river barge often pollutes land and water along the way when HAZARDOUS WASTES leak out.

ALTERNATIVES TO LANDFILLS

To cut the amount of materials sent to landfills, Americans are urged to recycle—in many communities it is required by law. Paper, which accounts for more than 45% of a landfill's volume, aluminum cans, and glass and plastic containers can be recycled into new products.

Americans are also encouraged to reduce the amount of garbage they produce by reusing materials whenever possible, such as cooking with a piece of aluminum foil more than once; recycling old clothing and furniture by giving them to the less fortunate; refusing to buy items that are overpackaged; and choosing products made only from, or packaged in, recycled materials.

Another way people can reduce the amount of garbage sent to landfills is by COMPOSTING. In nature, when a living thing dies and decays, its MINERALS enter the soil to be used as nutrients by other living things. Composting does the same thing, with human help. Organic matter like leftover food scraps, grass clippings, and leaves can be gathered together, kept moist, and allowed to decay. The end product can be used as fertilizer to aid new plant growth.

INCINERATION

The second most common method of garbage disposal is by burning, or incineration. It may seem like an alternative to landfills because it reduces waste by 70%, but burning creates vast quantities of ash that must be disposed of—usually in landfills. Much of the ash contains toxic material that can contaminate land and water if it leaches out of the landfill.

Other problems with incineration include the cost—up to $400 million per incinerator—and the fact that incinerators wear out quickly. Burning can pollute the air with acid gases, heavy metals, and DIOXIN, unless costly antipollution devices are used. The ENVIRONMENTAL PROTECTION AGENCY (EPA) regulates the use of municipal incinerators, under the CLEAN AIR ACT of 1990. The EPA proposes rules calling for the removal of hazardous materials, like lead batteries, before waste is incinerated to reduce toxic emissions and ashes.

An advantage of incineration is that heat from burning garbage can be used to create energy. By 1991, cities in the United States had 136 waste-to-energy (WTE) facilities in operation. [*See also* ACID RAIN; AIR POLLUTION; AIR POLLUTION CONTROL ACT; BIOMASS; BUREAU OF RECLAMATION; CLEAN WATER ACT; CO-GENERATION; CONSERVATION; CONTAINER DEPOSIT LEGISLATION; DECOMPOSERS; DECOMPOSITION; FRONTIER ETHIC; HAZARDOUS MATERIALS TRANSPORTATION ACT; HAZARDOUS SUBSTANCES ACT; HAZARDOUS WASTE MANAGEMENT; HEALTH AND DISEASE; HUMUS; INDUSTRIAL WASTE TREATMENT; LAND STEWARDSHIP; LAND USE; LEACHING; MEDICAL WASTE; METHANE; OCEAN DUMPING; POLLUTION; RECLAMATION ACT OF 1902; RECYCLING, REDUCING, REUSING; RUNOFF; SEWAGE; SOLID WASTE DISPOSAL ACT; SOLID WASTE INCINERATION; TOXIC SUBSTANCES CONTROL ACT; and TOXIC WASTE.]

◆ Cattle, dogs, and goats feed on garbage in the streets of Calcutta, India.

Gasohol

A FUEL for AUTOMOBILES that is made by blending gasoline with alcohol. During the mid-1970s and again during the 1980s, the United States experienced shortages of PETROLEUM imported from the Middle East. As a result of these shortages, the prices of petroleum products, especially gasoline, rose substantially. The rising prices, along with long lines at gasoline stations resulting from the shortages, prompted gasoline suppliers to search for fuels that could be used in place of gasoline. One fuel that received much attention was gasohol, formed by blending unleaded gasoline with either methanol or ethanol.

Gasohol is used extensively as an automobile fuel in several South American countries, including Brazil. However, it is not yet widely used in the United States. Despite this fact, gasohol has several qualities that make it a promising fuel for future use in the United States. For example, it can be used in most automobiles without changing any automotive parts, such as carburetors and fuel lines. In addition, by reducing the amount of petroleum needed to produce fuel, the use of gasohol can help make existing supplies of petroleum, a NONRENEWABLE RESOURCE, last longer. Another advantage of gasohol is that the alcohol portion of the fuel is made from crops such as wood, grains, and sugar PLANTS, all of which are RENEWABLE RESOURCES. A disadvantage of gasohol is that, like gasoline, it gives off waste products that are polluting to the ENVIRONMENT. [*See also* ALTERNATIVE ENERGY SOURCES; AIR POLLUTION; BIOMASS; CARBON DIOXIDE; CARBON MONOXIDE; CATALYTIC CONVERTERS; and OPEC.]

Gene

Part of a molecule of deoxyribonucleic acid (DNA) that carries information needed to produce a trait in an organism. A molecule of DNA is shaped like a twisted ladder, whose rungs are each made of two units called *bases* or *nucleotides*. One gene is a set of dozens, or hundreds, or even thousands of nucleotides, lined up along one half of the ladder shaped DNA molecule.

When a gene is to be used, the DNA molecule splits open like a zipper along the length of that

◆ Bacterial genes are located on these circles of DNA.

gene. Other molecules of a substance called *messenger RNA* then line up along the gene and copy its information, as if the string of nucleotides were a long word with many letters. Messenger RNA brings

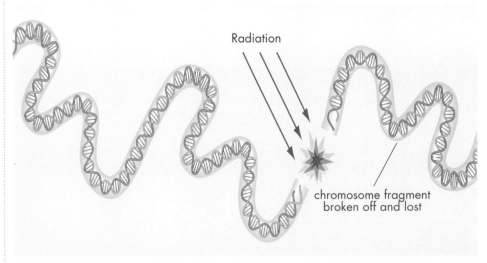

◆ A mutation can be caused by radiation. In this case, the mutant gene is missing a segment of itself.

Radiation

chromosome fragment broken off and lost

this copied information to another part of the cell, where the information from the gene is used to make proteins.

Making proteins is not the only activity directed by genes, but it is extremely important. We often think of proteins only as something that we eat, or something found in our muscles and hair. In fact, DNA has codes, or genes, for thousands of kinds of proteins that do a huge variety of jobs. They form all sorts of body tissues; help digest food; fight off disease; and also make it possible to move and think.

Proteins that serve as building materials in the body are called *structural proteins*. Proteins that assist in all the thousands of chemical reactions that keep the body alive are called *enzymes*. Other pro-

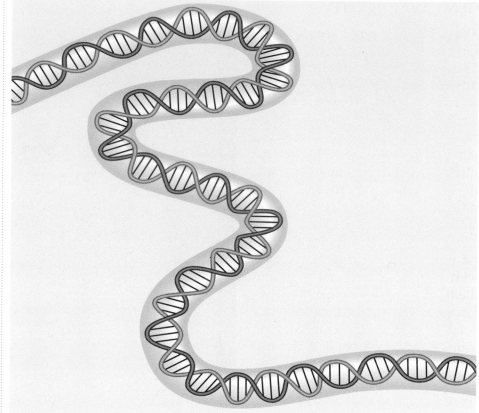

◆ A gene is a very long segment of DNA that is tens of thousands of units, or bases, in length.

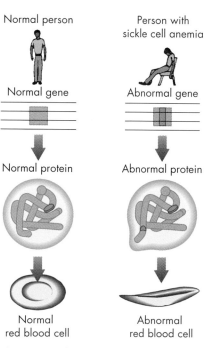

Normal person

Person with sickle cell anemia

Normal gene

Abnormal gene

Normal protein

Abnormal protein

Normal red blood cell

Abnormal red blood cell

◆ The genes of a normal person and a person with sickle cell anemia will have slight variations.

teins perform other jobs. Acting together, these proteins produce all the features, or traits, that make an organism what it is. Eye color, hair type, bone structure, blood type, and some skills and behaviors are all examples of traits passed on from parents in the genes.

Some traits can be changed or hidden by events in an organism's life; for example, a person who inherited the trait of tallness might remain short if kept on a very poor diet. Traits like this are said to depend partly on "nature" (the genes) and partly on "nurture" (the ENVIRONMENT in which the organism grew). Other traits, for example hair color, are present (or "expressed") no matter what the environment.

Many kinds of organisms have two of each gene because they inherit one copy from each parent. The two copies may be different. A person's two genes for eye color, for example, may be a gene for blue eyes and a gene for brown eyes. (In this case, both eyes will be brown, because that gene is dominant over the other one). In a whole population, however, there can be more than two types of the gene. The more types of each gene there are in a population, the greater the population's GENETIC DIVERSITY.

GENETIC DIVERSITY

The genes of a turtle are quite different from the genes of a daisy. These two SPECIES naturally need different instructions for growing and staying alive. However, within a species—a turtle species, for example—individuals have somewhat different sets of genes. This genetic diversity creates genetic variation in the population of turtles. They do not all inherit exactly the same color or size or shape. They do not all swim the same way. Some might be able to withstand the coldest WEATHER while others can withstand the hottest weather. Some may lay a few large eggs; some may lay many small eggs. This genetic variation would not be needed if the environment was always the same everywhere. However, in real life there are differences from place to place and from year to year, in a turtle pond or in almost any other HABITAT. If the turtles in this population were all exactly the same, then an event that was fatal to one turtle might kill them all. A single disease or change of weather might make the whole population extinct. If the turtles all have slightly different traits, then changes in the environment might not be equally bad for every turtle. A change might even be better for some of them. Thus, genetic variation sometimes protects a population, or even a whole species, from becoming extinct when conditions change.

Genetic Diversity and New Species

When organisms breed, copies of their genes are made and passed on

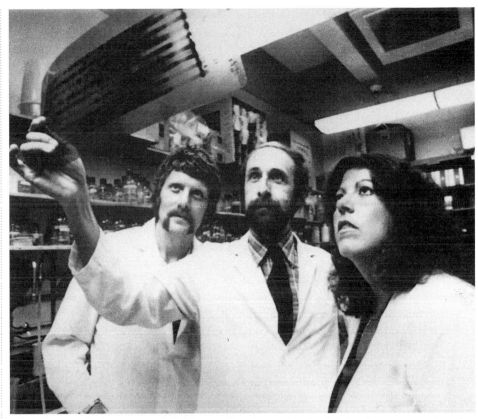

◆ Workers at the University of California–San Diego Medical School inspect copies of the human gene that causes a genetic disease.

to the next generation. Some of these copies are not exact. The new organisms may have a change, or *mutation*, in one or more genes. Some mutations have no effect; many have bad effects. A number of human diseases and birth defects are the result of gene mutations, some of which can be passed on from generation to generation.

Now and then, a mutation turns out to be an improvement. It gives an individual a new trait that helps it to survive and breed better than its parents. In very rare cases, a mutation may result in a new species. For example, a gene mutation might change the mating call in one family of an INSECT species. Mem-

bers of that family might breed only with each other because their new call did not attract any other member of the species. If their young continue to breed only among themselves, they may in time form a new species.

New species do not come entirely from gene mutations. Sometimes they come from other changes in the GENE POOL of a species. If, for example, a species of mouse lives in a FOREST, and a piece of forest habitat is changed to a field, this may be harmful to many of the mice. However, it may be good for a few forest mice who happen to have rare genes for grass-colored fur, for nesting in grass, and for other helpful traits. These mice might survive longer and have more young in the field than other mice do. Mice with the rare genes may soon become the most common kind of mouse in the field. Thus, in the new setting, the "rare" genes have become common genes through the process of NATURAL SELECTION. This sort of change in the population's genes is called a shift in *gene frequency*. Changes in gene frequency, like gene mutations, can contribute to the EVOLUTION of new species.

GENES AND THE ENVIRONMENT

Even though a certain amount of gene mutation is natural and even helpful, too much mutation can interfere with growth and health, cause CANCER, or cause birth defects in offspring. Certain chemical pollutants, including some of the heavy metals and some DIOXINS, can cause mutations, as do some forms of RADIATION. This is one reason why some HAZARDOUS WASTES cause concern.

Another concern is the loss of genetic diversity in small populations of ENDANGERED SPECIES. A group of only 100 people could not have every possible shape of human nose; in the same way, a group of 100 TIGERS, or 50 GRIZZLY BEARS, cannot carry all the different genes once found in larger populations of these species. When genetic diversity is lost, species are less able to survive changes, adjust to new environments, or evolve into new species. Efforts to save endangered species often include special breeding projects to save as many types

of genes as possible. [*See also* BIO-DIVERSITY; CARCINOGEN; EVOLUTION; GENE BANK; GENETIC ENGINEERING; and GENETICS.]

Gene Bank

▶ Any system involving the storage of an organism's GENES. Once a SPECIES becomes extinct, its genes are lost forever. Gene banks save PLANTS for agriculture. They also preserve specimens of rare or ENDANGERED SPECIES in case these organisms do become extinct.

Plants ar e typically stored as seeds. Animals are stored as frozen sperm and eggs. Because the genetic material is stored in refrigerated and humidity-controlled environments, genetic material can survive for many years.

Zoo organizations are working to preserve fertilized eggs from rare species, but the most common type of gene banks are seed banks. Today, more than 3 million seed samples are stored worldwide in about 100 seed banks. For example, the federal government maintains the National Seed Storage Laboratory in Fort Collins, Colorado, as part of its commitment to maintain BIODIVERSITY.

The task of preserving seeds is very large. Therefore, a number of

◆ The Royal Botanic Gardens in London, England, has been seedbanking since the 1960s. Today, it stores a variety of tropical plant species that may be lost through deforestation.

private organizations and research institutions worldwide have begun their own seed banks. Should plant species become extinct in the future, stored seeds could be used to reintroduce the species in suitable HABITATS. Seeds stored in seed banks can also be used to grow plants that can be crossbred to produce new plants with desirable traits. This technique can produce new species of crop plants that grow faster, contain more nutrients, or are resistant to disease or INSECTS.

One of the most successful seed banks is operated by the Royal Botanic Gardens in London, England. The Royal Botanic Gardens, also known as the Kew Gardens, began seedbanking in the 1960s. Today, the seedbank at Kew Gardens has stored more than 3,000 plant species. Currently, the Kew Gardens is collecting seeds from a variety of tropical plants in South and Central America. Its goal is to preserve the biodiversity that is rapidly being lost through DEFORESTATION. [*See also* ENDANGERED SPECIES ACT; GENETIC DIVERSITY; GENETIC ENGINEERING; HABITAT LOSS; SPECIES DIVERSITY; and WILDLIFE CONSERVATION.]

Gene Pool

▶ All the different GENES carried in a population. A gene pool is the sum of all the possible traits a certain SPECIES can have. Traits carried in the gene pool vary with each species of organism. For example, the human genes carry information

about hair color, eye color, skin color, height, and artistic ability, among other traits. All these genes and the others that determine the traits of humans make up the human gene pool. For horses, the gene pool would include the genes that determine mane color, hair color, markings, and speed. For lettuce, such traits as leaf color, taste, and kind of root are carried in the gene pool.

When the ENVIRONMENT changes, individuals having genes that give them the traits needed to adapt to these changes survive. Those members of a population that do not have the ability to adapt to changes in the environment do not survive. This process is called NATURAL SELECTION. Natural selection changes the gene pool. Because genes that promote survival are passed from generation to generation, natural selection is the way in which EVOLUTION, or the changing of life on Earth, takes place.

People who work to make sure that species of organisms do not become extinct are concerned about gene pools. Once a particular gene pool is gone, there is one less

◆ All of the genes in a population of any species, such as humans or caribou, make up the gene pool.

set of possibilities for a living organism. This loss diminishes life because all living organisms depend upon one another for the things they need to survive. [*See also* ADAPTATION; BIODIVERSITY; GENE BANK; GENETIC DIVERSITY; GENETIC ENGINEERING; GENETICS; and SPECIES DIVERSITY.]

Genetic Diversity

▌The variation of characteristics among members of a SPECIES. Variations such as the coloring of mature and immature individuals of males and females may indicate certain categories within a species. Other variations, such as the length of a mouse tail or the size of a rose petal, may be measurable. Slight differences in coloring or body size among all individuals are other kinds of variations. Variations, like characteristics, result from the GENES of an organism.

Genes are sets of information in the DNA of the cell that are passed from parent to offspring. Each individual gets a set of genes from each parent. The way these genes interact determines the characteristics of the individual. Sometimes, when the two parents' genes are combined, changes called *mutations* may occur. The genes of an individual may also be changed or mutated by factors in the ENVIRONMENT, such as RADIATION or chemicals. Mutated genes produce variations.

Variations can benefit an organism, harm it, or have no observable

◆ The fur color of the kittens in this family of cats varies among the kittens and from that of their mother.

effect. A PLANT may develop a variation of large flat leaves in the shade and scalloped leaves in the sun. The scalloped leaves, because of their reduced surface area, are less likely to be harmed by the sun's heat. Sometimes a variation can benefit the entire species. For example, some INSECTS hatch with a variation that makes them resistant to certain INSECTICIDES, while other insects are killed by it. The resistant insects survive to produce offspring that are also resistant to the insecticide.

Insects resistant to the insecticide are said to have adapted to their environment. The survival of the fittest, the rule of NATURAL SELECTION, has weeded out those individuals unable to adapt to their environment. The insecticide-resistant insects live to reproduce more insecticide-resistant individuals. Thus, the insect has changed or evolved. Natural selection is the process by which EVOLUTION takes place.

New knowledge about inheritance and genes is enabling scientists to find ways to produce useful variations in plants and animals. For the farmer, these new variations result in plants resistant to certain pests and diseases or plants that produce a greater yield. Correcting harmful variations in humans is the key to treating many serious illnesses. [*See also* ADAPTATION; BIODIVERSITY; GENE BANK; GENETIC ENGINEERING; and GENETICS.]

Genetic Engineering

▌The process of changing the inherited characteristics of a living organism. GENES are tiny structures in the nucleus of the cells of PLANTS, animals, and other organisms that

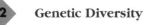

direct the inheritance of the individual. Genes are passed on by the individual's parents and control the physical appearance, structure, and functioning of the organism. Genetic engineering involves the insertion of new genes into the cells of an organism. Scientists have developed several ways for achieving this result.

In one procedure, a VIRUS is used as the transporter of the gene. Viruses are cell invaders by nature. Moving into the nucleus of the cell, the virus picks up a gene. Then the virus erupts from the cell, moving on to carry the gene to the nucleus of another cell. Scientists can mix the virus in a flask containing a particular gene, and later mix it with the cells of the organism for which the gene is intended. The virus will invade the nucleus of those cells, delivering the gene.

Another procedure, called *recombinant DNA*, uses tiny structures inside BACTERIA called *plasmids*. Plasmids are small rings of about ten genes. The plasmids are removed from the bacteria. Next,

using special chemicals called *enzymes*, scientists break the plasmid ring, insert a new gene, and put the plasmid ring back together. Then they mix the new plasmids with a flask of the cells that are in the cell walls to receive the new gene. A transfer of the flask from an ice bath to a warm bath causes the pores in the cell walls to open. The plasmids pour into the cells through the pores of the cell walls. The cells now have a new gene.

Another genetic engineering technique involves combining two cells to produce a hybrid cell. The hybrid cell possesses the genes of both cells, but the genes may not produce all the characteristics of the two parent cells.

Cloning is yet another genetic engineering technique. Normally an individual gets a different set of genes from each parent. Thus, the individual has a pair of genes for each characteristic. But they may not be exactly the same, since they come from two different individuals. For example, one gene for hair color may call for blond hair,

whereas the other calls for black hair. Working together, the genes may produce an individual with red hair. Other gene pairs in the individual will produce other characteristics different from one or both of the parents. Thus, no individual is exactly like another. Cloning, however, enables the scientist to produce an individual exactly like its parent. Genes from the cells of an adult mouse are inserted into a fertilized mouse egg from which the original genes have been removed. The egg then develops into a mouse that has the exact same genes as the parent mouse. The second mouse is a clone of the first mouse.

Genetic engineering bypasses inheritance and manipulates life to the scientist's purpose. Geneticists —scientists who work with genes— were the first to worry about the results of genetic engineering. In 1975, they raised fears that using plasmids to change genes in bacteria might create dangerous bacteria that could cause incurable infections or CANCER. The scientists had many meetings and conferences. One geneticist, Robert Sinsheimer, noted that in nature there are built-in barriers that prevent the mixing of genes, but scientists are breaking those barriers. How will this affect the balance of nature? he asked. Other geneticists said that gene exchange goes on all the time in nature. The rule is "anything goes."

The scientists agreed that rules were needed. They made strict rules for genetic engineering laboratories that were designed to regulate the kind of research to be done and to guard against experimental organisms being released from these laboratories. The results

◆ A genetically engineered, pest-resistant cotton boll (left) is compared to a normal cotton boll damaged by a cotton weevil.

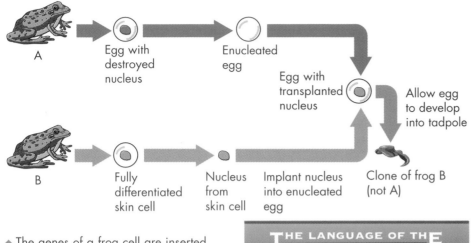

A   Egg with destroyed nucleus   Enucleated egg

Egg with transplanted nucleus   Allow egg to develop into tadpole

B   Fully differentiated skin cell   Nucleus from skin cell   Implant nucleus into enucleated egg   Clone of frog B (not A)

◆ The genes of a frog cell are inserted into a frog egg. The egg then develops into a clone of the frog—a tadpole that has exactly the same genes as the parent.

coming from those genetic laboratories are having and can have profound effects on life as we know it.

FLAWED GENES CAUSE ILLNESS

Genes regulate the way in which our bodies function. A faulty gene that fails to produce the proper chemical can make the individual severely ill. For example, the bodies of people with **cystic fibrosis** have a faulty gene that fails to prevent the buildup of sticky mucus in the lungs, intestines, and other parts of the body. Giving them a normal gene could make these people well. Such treatment may soon be available for people with cystic fibrosis. Other inherited diseases, as well as certain cancers, can be treated with gene therapy.

Genetic engineering can be used to make large quantities of medicines to fight infection and treat other diseases. Bacteria are given genes that enable them to make substances normally made by

the human body, such as insulin for diabetics, antibodies for infection, and interferon for use against virus infections and certain cancers.

Cloning might be used to grow a new body part, such as a leg or a heart. There would be no need to worry about **rejection** of strange tissue by the person's bodily defenses since the part would grow from the person's own cells. Cloning could also produce scores of champion racehorses from one individual. Animals on the brink of EXTINCTION could be cloned, too.

Recombinant DNA techniques have already created oil-eating bacteria that clean up oil spills. Other bacteria that readily break down SEWAGE have been created.

These are just a few examples of the broad horizons of genetic engineering. But geneticists pro-

---

◆ A virus enters the nucleus of one cell, removes a gene from that cell, and transfers it to the nucleus of another cell.

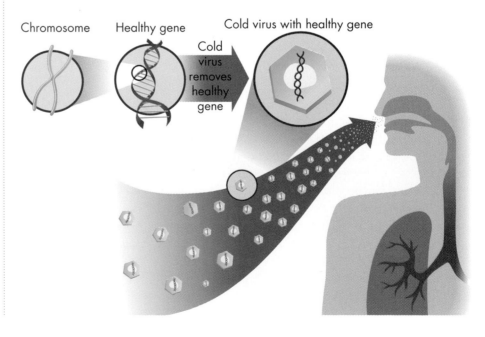

Chromosome   Healthy gene   Cold virus with healthy gene

Cold virus removes healthy gene

ceed carefully, studying the side effects of their work as they go. [*See also* DNA; EVOLUTION; GENE POOL; GENETICS; and HYBRIDIZATION.]

Genetics

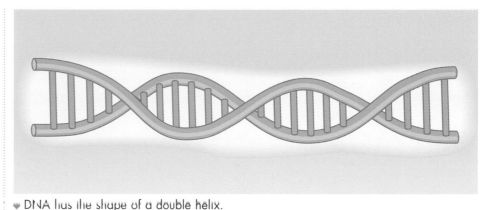

💗 DNA has the shape of a double helix.

▮The study of the way in which characteristics are passed from parent to offspring. GENES, the hereditary units, are in the nucleus of all the cells of organisms. The tiny structures are located on a larger, threadlike structure called a *chromosome*. Chromosomes come in pairs, with one gene for a given character on each chromosome. Every kind of organism has a certain number of chromosomes—humans have 46, or 23 pairs. Each chromosome has many genes on it. Humans are believed to have 50,000 to 100,000 genes.

Chromosomes are passed on from generation to generation during reproduction, the process of producing a new individual. Within the body, as new cells for repair or growth are made, the same set of chromosomes is passed on to the new cells. The reproductive cells—eggs and sperm—will pass on those chromosomes to offspring.

The genes on the chromosomes regulate the development of structure and the functioning of the individual. Genes do their job by directing the manufacture of certain proteins. For example, one gene may make a protein that forms a skin cell; another may make a protein that helps the individual to breathe. While each cell has the same genes, only those needed in the particular area function. The rest are turned off by the repressor or control genes. For example, only those genes that regulate eye structure and activity function in the eye.

THE GENE'S CHEMICAL CODE

The chemical makeup of the gene enables it to direct the body's activity. Genes are made up of a chemical called *deoxyribonucleic acid* (DNA). This chemical contains nitrogen bases that the body recognizes as a chemical code. A messenger chemical, *ribonucleic acid* (RNA), copies the code from the DNA in the nucleus and carries it to the outer part of the cell. There messenger RNA gathers the necessary ingredients and matches them to the pattern that messenger RNA has delivered. Every single protein made in the individual's body begins in this way. These proteins, in turn, direct the production of other chemicals made by the body. Thus, the manufacture of the skin on your finger, the fingernail, the blood vessel inside your finger began as RNA delivered the message from DNA deep inside the cells of your finger.

If even one of the nitrogen bases in the gene's chemical code is out of place, the wrong protein will be made. This could have serious consequences for the individual. For example, a protein that regulates digestion of food or the function of muscles could be missing. Such a flaw in a gene is called a *mutation*. A mutation may be passed on from a parent. A gene can also be damaged by a chemical, RADIATION, or even a VIRUS. In some cases, entire chromosomes may have twisted or reversed their positions as the parent cells joined. This can cause serious problems for the offspring, such as mental retardation.

A mutation can cause the death of an individual, thus preventing the individual from passing on the mutated gene to future generations. Some mutations, however, may prove beneficial in helping an individual survive in a particular ENVIRONMENT. That individual will live to pass on the beneficial mutation to future generations. This is called

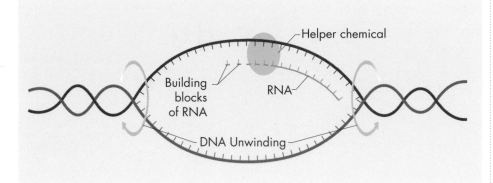

◆ Two strands of DNA separate, and a strand of RNA is copied from one DNA strand.

NATURAL SELECTION, or survival of the fittest.

An example of both a beneficial and a harmful mutation is the one that causes the development of an abnormal kind of **hemoglobin**, called *sickle* hemoglobin. This mutation developed in people living in warm climates in Africa and the Mediterranean area. Sickle hemoglobin is harmful to **malaria** parasites that invade red blood cells, thus protecting people having those cells from malaria. But there is a harmful side to this mutation. The person's red blood cells change from round to sickle shape when they are starved for OXYGEN or when the body's need for oxygen increases. Stresses such as cold WEATHER constrict blood vessels, causing what is called sickle cell crisis. The sickle cells clog blood vessels, killing many body cells, thereby damaging organs and tissues and causing a great deal of pain. There is no cure for this condition, but doctors do have ways to make the patient more comfortable.

Physicians are able to treat many of the mutations that produce illnesses in humans. Scientists are now beginning to find ways to replace mutated genes with normal ones. This is called GENETIC ENGINEERING. [*See also* DARWIN, CHARLES ROBERT; EVOLUTION; GENE; GENE POOL; GENETIC DIVERSITY; and NATURAL SELECTION.]

THE LANGUAGE OF THE ENVIRONMENT

malaria a tropical disease characterized by high fever, chills, and weakness. It is caused by a microscopic organism that invades the blood cells.

hemoglobin the protein in red blood cells that carries oxygen.

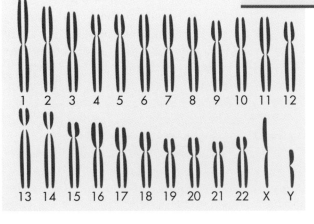

◆ The human has 23 pairs of chromosomes. The twenty-third pair determines the sex of the individual. Two X chromosomes produce a female. An X and a Y chromosome produce a male.

Geothermal Energy

◗ The useful energy that comes from the heat found in Earth's interior and brought to Earth's surface as hot water or steam. Heat from molten, or melted, rock warms water in locations where Earth's crust is thin, causing hot water and steam to rise to the surface and emerge through **geysers** and other natural vents. Such naturally heated water or steam can be used to power **turbogenerators**, creating electric power, or to directly heat buildings. Geothermal energy can also be tapped by drilling into hot rocks and pumping in cold water. As the water enters the ground around the molten rocks, it turns to steam and emerges, under pressure, from other openings.

Only a fraction of the total amount of geothermal energy is usable. However, the geothermal energy in the upper 3.1 miles (5 kilometers) of Earth's crust is estimated to be equivalent to 40 million times the energy contained in the world's supplies of crude oil and NATURAL GAS. Geothermal power now supplies one tenth of a percent of the world's energy requirements. It is used by nineteen countries, including Iceland (where it is the most important source of power), the United States, Italy, New Zealand, Nicaragua, Japan, China, the Philippines, and Kenya.

THE LANGUAGE OF THE ENVIRONMENT

geysers springs that emit boiling water and steam.

turbogenerators machines that produce electricity using a motor driven by the pressure of steam, air, water, or gas.

PROBLEMS WITH GEOTHERMAL ENERGY

Geothermal energy must be managed properly. Although geothermal heat is unlimited, the amount of groundwater that is depleted when steam is taken out is limited. In 1988, the largest facility in the United States, located approximately 68 miles (110 kilometers) north of San Francisco, produced around 2,000 megawatts of power, about equal to that produced by two large nuclear plants. This plant run by geothermal energy was projected to produce 3,000 megawatts by the year 2000. However, because steam was taken out faster than it could be replaced, the facility's output dropped to 1,500 megawatts by 1991. The expected power output from this plant for the end of the century is now projected to be just half the peak 1988 level.

Another problem with geothermal energy is POLLUTION. Hot water and steam from below ground sometimes contain salts, sulfur compounds, and other corrosive contaminants that are leached from MINERALS in the bedrock. When contaminated steam is released into the ATMOSPHERE, it can contribute to SULFUR DIOXIDE pollution. Hot water containing salts released into streams and rivers can be a source of both chemical and THERMAL WATER POLLUTION.

The possibility of HABITAT destruction is also a concern. At Pahoa in Hawaii, the construction of a geothermal plant on the side of an active volcano has been opposed. Some people claim the plant will lead to the destruction of large amounts of lowland tropical FOREST. Supporters of the plant claim that only 2% of the Wao Kele forests in the surrounding area is likely to be lost. [*See also* AIR POLLUTION; ALTERNATIVE ENERGY SOURCES; FOSSIL FUELS; FUELS; LEACHING; PETROLEUM; and THERMAL WATER POLLUTION.]

Giant Panda

▌A large bearlike animal whose natural HABITAT is the bamboo FORESTS in the mountains of central China and eastern Tibet. The giant panda is an ENDANGERED SPECIES. It is estimated that there are only 500 to 1,000 giant pandas left in the entire world, including those living in ZOOS.

The giant panda's most distinctive feature is its thick, wooly, black and white coat. It has black around its eyes and on its ears, legs, and chest. It also has a band of black across its shoulders. An adult giant panda grows to be about 5 to 5 1/2 feet (about 1.5 meters) from nose to tail and may weigh 165 to 330 pounds (75 to 150 kilograms) or more. It has a short, stubby tail.

Giant pandas are considered to be OMNIVORES because they sometimes eat small animals. However, most of their diet consists of bamboo, especially the shoots and roots. Because bamboo has little nutritional value, a giant panda must eat about 35 pounds (16 kilograms) of leaves and stems or about 90 pounds (40 kilograms) of shoots and roots each day.

THREATS TO GIANT PANDAS

The destruction of the bamboo forests threatens the survival of giant pandas. They will starve if

◆ To satisfy its energy requirements, a giant panda may eat from 10,000 to 35,000 pounds (about 4,500 to 16,000 kilograms) of bamboo a year.

the bamboo disappears. Bamboo PLANTS flower, produce seeds, and die in cycles of about 40 to 100 years. Thus, parts of bamboo forests sometimes die naturally, and the giant pandas must move on to other parts of the forest.

People may cut down entire forests to get the bamboo or to use the land for farming. When entire forests are destroyed, the pandas have no food. It may take six years before the forest can again sustain the pandas. Within the last 25 years, two entire bamboo forests were destroyed, resulting in the death of approximately 200 giant pandas. Another threat to the survival of giant pandas is HUNTING. They are sometimes hunted for their fur.

In an attempt to prevent the EXTINCTION of giant pandas, the Chinese government established 12 panda reserves with several captive-breeding centers. But the giant panda's mating habits are poorly understood, so captive breeding has not been very successful. [*See also* CAPTIVE PROPAGATION; HABITAT LOSS; POACHING; and WILDLIFE CONSERVATION.]

Gill Net

◗ Type of fishing net that is suspended vertically in the water to catch FISH by their **gills** as they attempt to swim through. Gill nets catch any fish that has a head small enough to fit through the **meshes** of the net, but a body that is too

◆ Fishers use gill nets to catch fish by their gills.

◆ These sockeye salmon have been caught by gill nets.

THE LANGUAGE OF THE ENVIRONMENT

gills respiratory organs found in fish and other aquatic organisms through which oxygen and water are obtained.

meshes openings between cords and threads of a net.

large to fit through them. Once the head of a fish passes through the mesh, the fish can no longer move forward. The fish cannot back out of the mesh because its gill covers become entangled in the mesh.

There may be environmental problems associated with the use of gill nets. For example, gill nets catch and kill fish of many SPECIES. Thus, both fish that are useful as food and those that are not can be caught in the nets. Another problem with the use of gill nets is that they continue to capture fish even if the nets become separated from fishing boats. One type of gill net, sometimes called a drift net, can be miles long. When these nets are torn loose from a boat, they may kill a great number of sea animals while drifting. [*See also* FISHING, COMMERCIAL.]

Glaciation

The EROSION of land and rock by the movement of thick sheets of ice called GLACIERS. Glaciers form in high latitudes where the amount of snowfall exceeds the rate at which the snow melts. When the glacier gets very thick, it begins a gradual downhill movement.

As it moves, a glacier sharpens mountain peaks and cuts deep grooves in rock. It moves huge rocks and great quantities of clay and gravel called *till*. The till is deposited along the path of the glacier, creating mounds and ridges of clay and rock called *moraine*. Till deposited alongside the glacier's path forms groups of long, low mounds called *drumlins*. The course of the Missouri River is enclosed by drumlins.

Streams of water produced by a melting glacier shape the till, creating plains and terraces. When huge chunks of melting ice are buried by meltwater deposits, deep holes or depressions in the plains are created.

Signs of the last great period of glaciation, which ended about 11,000 years ago, can be seen in many parts of the world. The numerous swamps and lakes in Canada, northern Europe, and the northern United States were formed in depressions created by glaciation. The Great Salt Lake in Utah was once a much larger glacial lake called Lake Bonneville. [*See also* EROSION; ICE AGE; and SEDIMENT.]

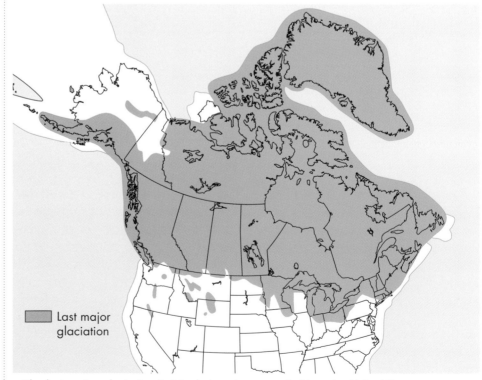

◆ The last major glaciation in North America extended into the United States.

◆ More than four-fifths of the island of Greenland is covered with ice. The average thickness of the ice cap is one mile.

Glacier

▌A moving mass of ice that forms from layers of compressed snow. Over time the ice forming a glacier becomes so thick that the glacier begins to move across the land under the pressure of its own weight.

Since about 20,000 B.C., Earth's CLIMATE has fluctuated between warmer and cooler periods. These periods, which lasted about 10,000 years, were sandwiched between major ICE AGES that lasted about 100,000 years.

Glaciers begin forming in places that receive more snow in winter than can melt or evaporate in summer. Over the years, the excess snow gradually builds up in layers. The weight of upper layers presses down on the snow crystals beneath. The crystals are compressed into a dense ice called *firn*, or snow that has passed through one summer melt season but is not yet glacial ice. When the firn becomes greater than 50 feet (15 meters) thick, it is glacial ice. The depth of most glaciers is between 300 and 10,000 feet (90 and 3,000 meters).

KINDS OF GLACIERS

There are two main types of glaciers: *alpine glaciers* and *ice sheets*. Ice sheets are sometimes called *continental glaciers*. The two types of glaciers differ in size, shape, and location.

Alpine glaciers form on mountains, sometimes in bowl-shaped depressions between peaks. These glaciers move down mountain slopes through valleys. One of Switzerland's largest alpine glaciers

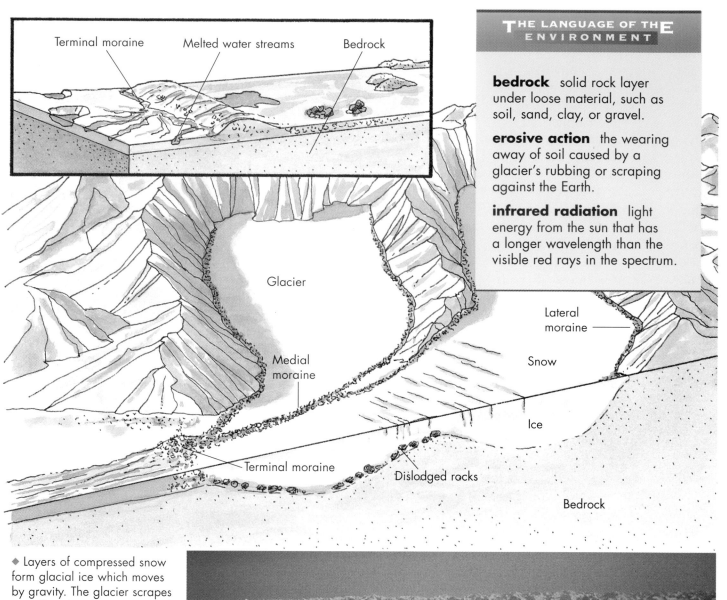

Terminal moraine   Melted water streams   Bedrock

Glacier

Medial moraine

Lateral moraine

Snow

Ice

Terminal moraine

Dislodged rocks

Bedrock

◆ Layers of compressed snow form glacial ice which moves by gravity. The glacier scrapes up soil and rock, moves it and deposits it at the glacial edge. The deposits are called *moraines*.

◆ As alpine glaciers (also called mountain glaciers) move down the sides of mountains, it is easy to see why they are called "rivers of ice."

is the 9-mile long (15-kilometer long) *Gornergletscher*, or Corner Glacier, near the town of Zermatt.

Ice sheets, or continental glaciers, build up broad center domes and spread in all directions toward the sea. They cover huge areas of land, including valleys, plateaus, and mountains. Today, continental ice sheets cover most of Greenland and ANTARCTICA. During the last major ice age (in the Pleistocene period), some 18,000–20,000 years ago, nearly one-third of Earth's land was covered by glaciers. Massive ice sheets covering most of North America and Europe carved rugged landscapes as they moved across the regions. The jagged peaks of Europe's Alps and California's Yosemite Valley are products of glacial movement. When glaciers reach the sea, huge pieces break off and float away as icebergs.

MOVEMENT OF GLACIERS

Glaciers move because of the pull of gravity. Ice crystals at the bottom of the glacier slip over one another as they are squeezed by the layers above. These tiny movements within the glacier cause the whole icy mass to move. The movement is assisted by the melting of ice crystals at the bottom of the glacier. The melting results from the heat of friction and from heat given off by Earth.

Different glacier parts move at different speeds. Upper and middle layers flow the fastest. Bottoms and sides move more slowly because they rub against soil and rock walls. This difference in movement creates tension in the rigid, brittle uppermost part of the glacier, which fractures and forms deep crevices. Glaciers normally move less than 1 foot (30 centimeters) per day. However, some glaciers, called *galloping glaciers*, have been known to move more than 160 feet (50 meters) in a day. Scientists measure glacial movement by driving in stakes at varied spots and recording any changes in their position.

CHANGING EARTH'S SURFACE

As they melt and refreeze, the bases of moving glaciers scoop up and drag along large amounts of rocks and boulders. This rocky debris grinds and rakes loose rock fragments across the **bedrock** beneath, creating craggy countryside. The **erosive action** of alpine glaciers moving to valleys below often forms rounded hollows called *cirques* or scrapes out fjords— U-shaped valleys below sea level that become flooded by the sea. One of the best-known fjords, Norway's Sognefjorden, is more than 100 miles (160 kilometers) long, with walls that tower as much as 3,280 feet (1,000 meters) above and below the water.

Glaciers do not only carry away land; they also deposit debris. Piles of rock, dirt, and gravel dumped at the end of a glacier are called *terminal moraines*. Many forested hills are formed from terminal moraines. Materials dumped from a glacier's base as it retreats are called *ground*

◆ Glaciers, such as this one in Alaska, cause soil erosion.

◆ This is the end of a typical alpine glacier as viewed from the air.

moraines. Some fertile soil in the midwestern farm belt of the United States was formed from layers of moraines left by ancient ice sheets.

TEMPERATURES AND GLACIERS

Most glaciers grow slightly larger in winter, when cold temperatures assure snowy buildup and hinder the melting of lower layers. Glaciers can get smaller in summer, when rising temperatures cause lower layers to melt faster. Glaciers also increase or decrease in size as a result of CLIMATE CHANGES that occur over long periods of time. For example, Greenland's ice sheet has been gradually shrinking because the temperature there has been rising gradually since the early 1900s.

In Earth's ATMOSPHERE, CARBON DIOXIDE helps to warm Earth by preventing **infrared radiation** from escaping. Some scientists predict that a buildup of carbon dioxide in the atmosphere, resulting from the use of FOSSIL FUELS and other activities, may increase Earth's temperature. An increase of 8° F (4.5° C) in world temperature would make Earth's climate warmer than at any time in the past 100,000 years. Computer analyses suggest that such a temperature difference could melt glaciers and ice sheets. The water released from the melting glaciers could raise sea level as much as 230 feet (70 meters), leaving many coastlines under water. On the other hand, a drop in world temperature of 4° F (−16° C) could signal the beginning of a new ice age. [*See also* CLIMATE; ECOSYSTEM; EROSION; GLOBAL WARMING; GREENHOUSE EFFECT; and SEDIMENT.]

Global 2000 Report, The

▶ Study issued in 1980 under the joint direction of the U.S. State Department and the President's COUNCIL ON ENVIRONMENTAL QUALITY, forecasting the condition of the ENVIRONMENT by the year 2000. President Jimmy Carter created a task force to develop plans for preventing problems predicted in the report, such as the EXTINCTION of hundreds of thousands of PLANT and animal SPECIES, possible GLOBAL WARMING from growing amounts of CARBON DIOXIDE in the ATMOSPHERE, and an overpopulated world where people must compete for food and for water.

SOME WARNINGS

The task force concluded that there would be about 2.5 billion more people on Earth in the year 2000 than in 1975. Most of the POPULATION GROWTH would be in poor developing countries where food is already scarce. Millions of children would starve to death. Those who sur-

vived might have physical and mental handicaps.

The report also predicted that wealthy nations would have sufficient energy from conventional FOSSIL FUELS and ALTERNATIVE ENERGY SOURCES, but that poorer nations would have serious energy shortages. The 25% of the population dependent on wood for FUEL would cut so many trees for firewood that the deforested ground would erode and lose its ability to hold water, causing water shortages in many parts of the world. This would lead to the land becoming drier and to the loss of TOPSOIL, so important for growing crops.

In addition, a predicted 500,000 to 2 million plant and animal species would become extinct due to loss of HABITAT and polluted PRECIPITATION. But, the report added, these disasters could be avoided if all nations conserved fuel, protected animals and plants, cleaned the air and water, and worked to increase food supplies. [*See also* AIR POLLUTION; CLEAR-CUTTING; DEFORESTATION; ENDANGERED SPECIES; ENERGY EFFICIENCY; EROSION; FUEL WOOD; INFANT MORTALITY; SAFE DRINKING WATER ACT; SPECIES DIVERSITY; and WATER POLLUTION.]

Global Warming

▶ A hypothesis suggesting that an increase in worldwide temperatures will result from an increase in the amounts of CARBON DIOXIDE and other GREENHOUSE GASES in the ATMOSPHERE.

Scientists have observed that since the last ICE AGE, about 10,000 years ago, the CLIMATE in many regions of Earth has changed dramatically. One of the most significant changes has been a general increase in temperatures worldwide.

THE GREENHOUSE EFFECT

Carbon dioxide in the atmosphere helps to regulate temperatures on Earth. It does this by preventing heat energy near Earth's surface from escaping back into space. This action of carbon dioxide is called the GREENHOUSE EFFECT, because it is similar to how a greenhouse holds in heat energy from the sun.

SOURCES OF CARBON DIOXIDE IN THE ENVIRONMENT

Carbon dioxide, along with OXYGEN, is naturally cycled through the ENVIRONMENT, largely through the processes of PHOTOSYNTHESIS, RESPIRATION, and DECOMPOSITION. Through this cycling process, amounts of carbon dioxide and oxygen in the atmosphere are maintained at fairly constant levels. However, other processes increase the amounts of carbon dioxide released into the atmosphere. Among these processes is the burning of FUELS, especially FOSSIL FUELS such as COAL, PETROLEUM or oil, and NATURAL GAS. These fuels, especially coal and oil, are burned largely in power plants for the production of ELECTRICITY. However, they are also used to provide heat and cooking fuels in homes. In addition, gasoline—a petroleum product—provides the power for many

machines, including AUTOMOBILES, boats, buses, lawn mowers, and airplanes. All of these uses of fossil fuels and their by-products increase the amount of carbon dioxide released into the atmosphere.

Another source of increased carbon dioxide in the atmosphere is DEFORESTATION. PLANTS use carbon dioxide for photosynthesis. Some scientists are concerned that the cutting down of large numbers of trees in an area will increase carbon dioxide levels and lower oxygen levels by altering the amount of these gases involved in the CARBON CYCLE and OXYGEN CYCLE.

OTHER GREENHOUSE GASES

Because of its involvement in the greenhouse effect, carbon dioxide is called a greenhouse gas. However, carbon dioxide is not the only gas involved in the greenhouse effect. Other gases involved in this process include METHANE; chlorofluorocarbons (CFCS); and NITROGEN OXIDE. Like carbon dioxide, the amounts of these gases in the environment have also been increasing.

Methane is a gas that is sometimes used as a fuel. It is given off naturally as a result of decomposition from WETLANDS and rice fields. Studies have shown that cattle and other LIVESTOCK also give off significant amounts of methane as a result of their digestive processes.

◆ Fossil fuels burned in power plants to produce electricity increase carbon dioxide levels in the atmosphere and may contribute to global warming.

CFCs are present in some foam products such as polystyrene foam and solvents. CFCs were also once commonly used as propellants for AEROSOLS or as coolants in refrigerators and air conditioners. However, CFCs were shown to have harmful effects on Earth's OZONE LAYER. As a result, their use is being eliminated in aerosols and coolants.

Like carbon dioxide, nitrogen oxide is associated with fossil fuel use. It is also released into the environment through the use of fertilizers.

EVIDENCE OF GLOBAL WARMING

Scientists have done many studies to see how levels of carbon dioxide and other greenhouse gases have changed over time. They have conducted similar studies to see how Earth's temperature has changed over the same period. When the results of these studies are looked at together and graphed, it becomes clear that changes in global temperatures appear to match changes observed in carbon dioxide levels.

EFFECTS OF GLOBAL WARMING

Many scientists are concerned about how global warming might affect Earth. For example, scientists are certain that global warming affects the WEATHER patterns around Earth. If the temperature of the atmosphere increases, the temperature of Earth's OCEANS, which are involved in various weather conditions, will also increase. As ocean waters become warmer, severe

storms, such as hurricanes, would be likely to become more common. In addition, many areas of the world would experience changes in their rainfall patterns. Such changes would affect agriculture by changing the types of crops that can be grown in an area.

Another aspect of global warming that is of concern to many scientists is how this trend might affect the sea level. As ocean water and atmospheric temperatures rise, many scientists believe sea level will also rise because GLACIERS, ice caps, and icebergs in the far northern and southern parts of the world will begin to melt. These scientists believe that as these masses of frozen water melt, sea levels will rise, causing severe flooding in many coastal regions.

CONCERNS ABOUT GLOBAL WARMING

Scientists do not all agree on the human impact on global warming.

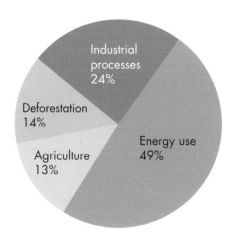

◆ The four major activities producing greenhouse gases are agriculture, energy use, industrial processes, and deforestation. Energy use, by far, is the greatest producer.

Many scientists believe the activities of people are affecting global warming. Others, however, believe that global warming is a natural process similar to those processes that have caused ice ages of the past. Although not all scientists agree that global warming is a major issue for future generations, those who are concerned have suggested some ways people can change their activities to help reduce the levels of greenhouse gases in the atmosphere and slow the global warming process. Several of these methods are:

- Use ALTERNATIVE ENERGY SOURCES in place of fossil fuels to meet the energy demands of a growing world population.

- Reduce deforestation.

- Plant trees and shrubbery in areas that have been cleared for use as homes or parks. Just as deforestation may increase the level of carbon dioxide in the atmosphere, planting trees and shrubbery in an area can help decrease carbon dioxide levels.

- Find alternative materials for use in polystyrene foam and solvents to reduce the amounts of CFCs released into the environment when items made from these materials are discarded.

- Reduce the amount of fertilizers used to grow crops. This will help limit the amount of nitrogen oxide that enters the atmosphere. [See also AGRICULTURAL POLLUTION; AIR POLLUTION; BIOGEOCHEMICAL CYCLE; BIOMASS; CARBON; CHEMICAL CYCLES; CONSERVATION; COUNCIL ON ENVIRONMENTAL QUALITY; ENVIRONMENTAL PROTECTION AGENCY (EPA); FUEL WOOD; GLACIATION;

HYDROCARBON; INDUSTRIAL REVOLUTON; LAW, ENVIRONMENTAL; METEOROLOGY; MONTREAL PROTOCOL; NATURAL DISASTERS; NITROGEN CYCLE; OCEAN CURRENTS; OZONE; OZONE HOLE; PETROCHEMICAL; PLANT; PRECIPITATION; SYNTHETIC FUEL; and ULTRAVIOLET RADIATION.]

Gorilla

▶Large-bodied African ape SPECIES that includes the mountain gorillas of Rwanda and eastern Zaire and the eastern and western lowland gorillas of central Africa. Gorillas are extremely powerful creatures and are the largest of the living primates. Male gorillas average more than 440 pounds (200 kilograms) in body weight. Females are much smaller than males, averaging about 200 pounds (90 kilograms). Along with their cousins, the chimpanzees, gorillas are the closest living relations to humans and are an ENDANGERED SPECIES.

CHARACTERISTICS OF GORILLAS

All gorillas have long, heavily muscled arms with grasping hands, a short trunk, and short powerful legs. Gorillas move slowly on the ground by knuckle-walking. In contrast, young gorillas love to climb trees.

Gorillas are also among the most gentle and social of all primates. Gorillas are HERBIVORES that

feed on the leaves, fruit, shoots, and stems of PLANTS. Gorillas have huge back teeth with which they grind their plant foods. Male gorillas also have a bony ridge along the top of their skulls that anchors thick jaw muscles. During feeding, gorillas use their powerful hands and arms to uproot young trees and strip them of leaves.

Gorillas live in family groups of about 10 to 12 individuals. There have been some reports of larger groups consisting of 30 individuals. These groups usually have one mature adult male known as the *silverback*, one or more younger males, and several adult females with their young. This group ar-

rangement is similar to that of many other primate groups. However, in gorilla society, females may move back and forth between groups. As a result, gorilla groups may be made up of unrelated females that do not interact with each other.

Most interactions in a gorilla society are between the dominant silverback male and individual females. The silverback usually has sole mating rights in the group. In this way, gorillas are similar to other primates, and there is often intense COMPETITION between males for control of the group. To avoid competition, maturing gorilla males may leave a group to look for females with which to mate.

The **home range** of the gorilla is 9 to 14 square miles (25 to 40 square kilometers). However, gorillas travel only about 1,640 feet (500 meters) each day in search of food.

◆ Gorillas, like many other primates, live in family groups.

◆ Mountain gorillas are threatened with extinction due to habitat loss and illegal hunting.

At night, gorillas build nests in trees by bending flexible branches and then sitting on them to form a platform. During the day, the gorillas build day nests in which they nap. Females share their nests with offspring.

THREATS TO GORILLAS

Gorillas have been intensely studied by scientists George Schaller and, more recently, the late Dian FOSSEY. Her book, *Gorillas in the Mist,* and the film based on it, documented the **natural history** of mountain gorillas and focused world attention on efforts made to protect gorillas from their only nat-

ural enemies—humans. Today, all SUBSPECIES of gorillas are threatened by DEFORESTATION, HABITAT LOSS, and illegal HUNTING, called POACHING. Live gorillas are often captured and traded illegally to ZOOS. In addition, many gorillas are killed for their skin and body parts, which are valued as trophies.

Due to the CONSERVATION efforts of Dian Fossey and others, mountain gorillas living in the Virunga Volcanoes National Park in Rwanda are somewhat protected. NATIONAL PARKS in Uganda and Zaire have also been established to help protect gorillas. However, fewer than 600 mountain gorillas now live in the wild. [*See also* BIODIVERSITY.]

Grasslands

❱ An ENVIRONMENT dominated by grasses and similar PLANTS. Grasslands are found worldwide in regions where PRECIPITATION is between 10 and 30 inches (25 and 75 centimeters) yearly.

Grasslands called PRAIRIES make up much of the central part of the United States from west of the Mississippi River to the foothills of the Rocky Mountains.

Springtime in grasslands is typically warm and wet. Grasslands also experience a dry season where rainfall is uncertain. The pattern and amount of rainfall in grasslands affects the quality of the SOIL, as well as the types of animal and plant SPECIES that can live there. For example, because there is insufficient water, grasslands contain few trees. Thus, although there is too much rain to support a DESERT environment, there is too little rain to support a FOREST.

FERTILE SOIL IN THE GRASSLANDS

In temperate grasslands, such as those in Canada, Europe, and the United States, soils are very fertile and rich in HUMUS. Grass plants die off during the harsh winter months, and decay into nutrients that build up in the soil. The soils and CLIMATE of grasslands are ideal for growing cereal grains such as oats, rye, wheat, and other species of grasses. For these reasons, grasslands are sometimes known as the "breadbaskets of the world." Other plant

◆ Drought-resistant wildflowers such as sunflowers thrive in grassland environments.

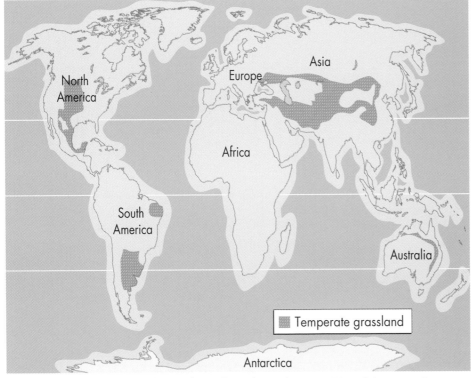

Temperate grassland

◆ Temperate grasslands are located mainly in the midwestern United States and in Asia.

species common in grassland environments include drought-resistant species of wildflowers, such as sunflowers and blazing stars.

ANIMAL LIFE IN THE GRASSLANDS

Although inadequate rainfall is the principal reason for the lack of trees in grassland environments, the animals that roam grasslands also keep trees from spreading. Vast herds of bison, zebras, antelopes, ELEPHANTS, and other HERBIVORES eat grass and other herbaceous plants and the new shoots of tree species before the sprouts grow too large. These large groups of GRAZING and browsing animals, in turn, attract many meat-eating animals, or CARNIVORES, such as the fast-running cheetahs and lions of the African savannas. The keen-eyed owls, hawks, and eagles of the United States are also carnivores that prey on many of the smaller herbivores, such as mice and prairie dogs.

Grasslands are also home to a variety of small animals, such as rodents, lizards, frogs, snakes, and INSECTS. To avoid being seen by PREDATORS, many of these species make their home underground. The common prairie dogs of the central United States, for example, have been known to construct extensive underground networks of tunnels.

IMPORTANCE TO HUMANS

Grasslands, especially those in the United States, are important to humans because of their use in agriculture. However, some agricultural and grazing practices have

◆ The Osage Prairie National Area in Missouri has been set aside for preservation.

◆ Most grasslands in the United States are being used for agricultural purposes.

damaged large parts of the grassland ENVIRONMENT. OVERGRAZING and trampling by herds of cattle and horses have killed much of the grass. Much of the rich TOPSOIL has also been lost due to water and wind EROSION of plowed land. Erosion caused by agricultural and grazing practices is a worldwide problem. Constant plowing and lack of rain changes the rich topsoil into dry, loose dust. In the 1930s, the central United States experienced a severe drought and became known as the DUST BOWL. High winds kicked up the dry soil and created a cloud of dust that covered much of the area and made growing crops impossible. The tragedy of the dust bowl resulted in new IRRIGATION and CONSERVATION techniques in grasslands. [*See also* AGROECOLOGY; BIOGEOCHEMICAL CYCLES; BIOME; ECOSYSTEM; and LAND USE.]

◆ Many grasslands in the United States are used for grazing cattle.

◆ In the United States, grazing on public lands is managed to prevent overgrazing, which can lead to environmental damage.

Grazing

❿ The eating of grass by such HERBIVORES as cattle, sheep, horses, and goats. Grazing occurs on RANGELANDS, which are vast open areas of land dominated by grasses and shrubs. Grazing is an important component of ranching.

When animals are allowed to graze in an area for too long, the area is said to be *overgrazed*. In an overgrazed area, the grass PLANTS are damaged by the constant movement of animals and vehicles. The root systems of the plants can prevent erosion. Since root systems help hold the soil together, OVERGRAZING leaves an area prone to soil EROSION.

If an overgrazed area also experiences long droughts, the land can become so damaged that it turns into DESERT. This process is called DESERTIFICATION. Desertification is a major problem in many developing nations, where environmental laws are often less strict than those in the United States.

In some parts of the world, such as the tropical regions of Africa and South America, rangeland is less available than it is in the United States. In the TROPICS, workers must clear vast stretches of RAIN FOREST to provide grazing areas for LIVESTOCK. In South and Central America alone, over 20,000 square miles (52,000 kilometers) of rain forest are cleared each year for cattle ranching.

In the United States, ranching is permitted on many public GRASSLANDS, as well as on privately owned property. Livestock graze on the NATIONAL WILDLIFE REFUGES, National Resource Lands, and NATIONAL FORESTS. The U.S. DEPARTMENT OF THE INTERIOR and DEPARTMENT OF AGRICULTURE (USDA) manage grazing on the public grasslands.

In 1978, the U.S. Congress passed the Public Rangelands Improvement Act, which helps protect public grasslands from overgrazing. Under this law, animal herds must be limited to sizes that the land can support. Herds must be moved often to prevent overgrazing in one area. Ranchers must maintain grazing lands by planting vegetation where soil is bare, building fences around overgrazed areas to let plants recover, and digging more water holes so that herds do not overgraze around a single watering hole. [See also DEFORESTATION; LAND USE; MULTIPLE USE; NATIONAL GRASSLAND; and PUBLIC LAND.]

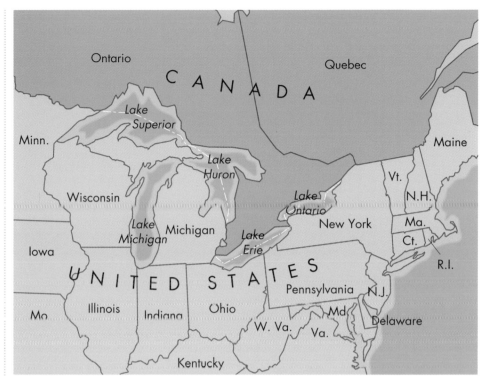

◆ The Great Lakes are on the boundary of the United States and Canada, in great hollows carved out of bedrock by glaciers. Both the United States and Canada discharge pollutants into the Lakes.

Great Lakes

▌The five lakes known as the Laurentian Great Lakes that are located between the United States and Canada. These lakes include: Lake Superior, Lake Michigan, Lake Huron, Lake Erie, and Lake Ontario.

The Laurentian Great Lakes have had their present shape and drainage outlets for about 2,500 years. They hold 20% of the world's fresh SURFACE WATER. The Great Lakes are the largest body of fresh water in the world. Today, about 37 million people live either along their 16,000 miles (25,600 kilometers) of shoreline or within their WATERSHED. The Great Lakes provide much of the water needed for steel production, inland shipping, commercial and recreational fishing, and municipal water supplies in the United States and Canada. In watershed areas of the Great Lakes, some water also is used to generate HYDROELECTRIC POWER.

Only 300 years ago, the Great Lakes were almost untouched by human development. Lake HABITATS varied from the cold, deep waters of Lake Superior to the warmer and shallower waters of Lakes Erie and Ontario. FISH were abundant, and some SPECIES grew to great sizes, such as sturgeon, which can reach lengths of 9 feet (3 meters) and weigh 400 pounds (180 kilograms). At this time, the lake shores were forested.

◆ One effect of the bioaccumulation of toxins in the Great Lakes: this cormorant, a fish-eating bird, has developed a severely deformed bill.

THE LANGUAGE OF THE ENVIRONMENT

agricultural related to farming and products produced on the farm.

overharvested condition in which a food source is removed faster than it can be replaced.

phosphorus a chemical used to make fertilizers.

Since European settlement began in the early 1600s, the Great Lakes have been subjected to ever-increasing human use. Many of the original fish species supported by the lakes have been **overharvested**. Thus, the lakes no longer support a large commercial fishing industry. SEWAGE and **agricultural** RUNOFF have disrupted the lake ECOSYSTEMS by overloading them with nutrients that cause explosions of ALGAE known as ALGAL BLOOM. PESTICIDES and toxins such as MERCURY and polychlorinated biphenyls (PCBS) that have been produced by industrial processes have entered the lakes' FOOD CHAINS, threatening the health of animals and people who consume lake fish. In addition, new species have been introduced into the lakes, causing dramatic changes in the population. Among the most notorious of these are the lamprey, an eel-like fish that kills other fish by sucking blood and fluids from them, and the zebra mussel, which takes over the food and living space needed by other species and also clogs water pipes.

Since the 1960s, much has been done by government agencies and citizens' groups in the United States and Canada to try to solve some of the environmental problems of the Great Lakes. An important role has been played by the International Joint Commission, a Canadian and American organization founded in 1909 to deal with international issues involving the Great Lakes. The Great Lakes Water Quality Agreements of 1972 and 1978 were joint Canadian-American efforts to control discharge of sewage and toxic pollutants. The CLEAN AIR ACT of 1990 was also helpful, since airborne pollutants affect the lakes.

Some problems in the Great Lakes have been effectively solved. For example, the POLLUTION caused by **phosphorus** in Lake Erie has largely been cleaned up, making this once severely damaged ecosystem a thriving BIOLOGICAL COMMUNITY. The Great Lakes will never be restored to their original condition. But efforts are made to improve lake quality. These efforts include phasing out the discharge of toxins, treating contaminated areas of lake bottom, and stabilizing the populations of any new species. [*See also* AGRICULTURAL POLLUTION; BIOREMEDIATION; EXOTIC SPECIES; FISHING, RECREATIONAL; NATIVE SPECIES; and PARASITISM.]

Greenhouse Effect

▶ The warming of the ATMOSPHERE that results from an increase in its levels of CARBON DIOXIDE and other gases. Earth is heated by the sun. Some of the heat that reaches Earth is reflected back into the atmosphere. Gases in the atmosphere, such as carbon dioxide, prevent this heat from escaping into space. Instead, the heat is held in and warms the atmosphere. This movement of heat between Earth's surface and the atmosphere helps regulate temperatures on Earth.

The gases in the atmosphere that help regulate temperature are called GREENHOUSE GASES. Greenhouse gases got their name because they work in much the same way as a greenhouse. The gases allow visible light to pass through them. However, they prevent the escape of infrared RADIATION, or heat.

CARBON DIOXIDE

Much of the carbon dioxide in the atmosphere is produced through natural processes. These processes include the RESPIRATION of organisms, the DECOMPOSITION of plant and animal remains, and the chemical breakdown of matter. Volcanic eruptions also release great amounts of carbon dioxide into the air.

More than 100 years ago, the development of the engine created a new source of carbon dioxide. Engines are powered by the burning of FUELS such as oil, gasoline, wood, PEAT, methanol, GASOHOL, and COAL. When such fuels are burned, carbon dioxide and water are given off as waste products. The carbon dioxide produced from the burning of such fuels is believed to have increased the carbon dioxide level of the atmosphere by one-third, from 260 to 345 ppm (parts per million). Scientists have observed that carbon dioxide in the atmosphere is increasing by 1.5 ppm per year. Since 1950, the greatest amount of this carbon dioxide is believed to be from the burning of FOSSIL FUELS.

OTHER GREENHOUSE GASES

Chlorofluorocarbons (CFCS), METHANE, and NITROGEN OXIDES are other greenhouse gases. CFCs are compounds of chlorine, fluorine, and CARBON. These compounds have been used in the production of polystyrene foam, as propellants in AEROSOLS, and as coolants in refrigerators and air conditioners.

Methane is a HYDROCARBON that is released into the atmosphere from flooded fields such as rice paddies, the digestive tracts of HERBIVORES, LANDFILLS, coal mines, forest clearing, and leakage from natural gas pipelines. Nitrogen oxides, compounds of nitrogen and oxygen, are produced by the burning of coal and the decay of fertilizers spread over farm fields.

GLOBAL WARMING

In the last century, Earth's average temperature has increased by 1.2° F (0.7° C). Some scientists predict the temperature will rise another 0.5° F (0.3° C) every ten years. A doubling of the amount of carbon dioxide present in the atmosphere would raise Earth's average temperature between 6.3° F and 8.1° F (3.5° C and 4.5° C). This possible increase in worldwide temperatures is called GLOBAL WARMING.

The results of global warming could drastically change Earth. For example, polar ice would melt,

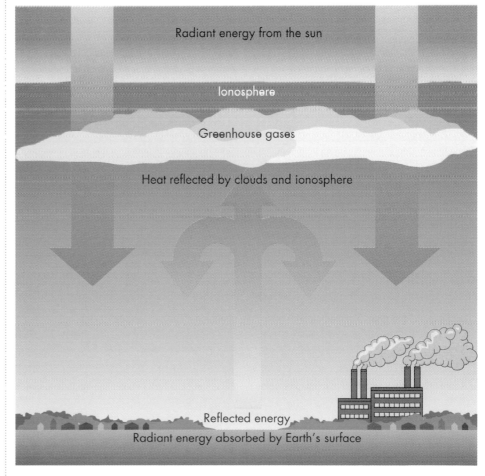

Radiant energy from the sun

Ionosphere

Greenhouse gases

Heat reflected by clouds and ionosphere

Reflected energy

Radiant energy absorbed by Earth's surface

◆ Greenhouse gases prevent reflected energy from leaving the atmosphere.

causing a rise in the water level of the OCEANS. A rise in temperatures would also cause changes in OCEAN CURRENTS. These changes, in turn, could alter CLIMATE patterns. The lakes and rivers that dot land ENVIRONMENTS would evaporate faster. All of these changes would alter the HABITATS of organisms. Those organisms unable to adapt to these rapid changes may not be able to survive in their changed habitats, resulting in MASS EXTINCTIONS.

SCIENTIFIC STUDIES OF GLOBAL WARMING

At a meeting held in 1985 by the World Meteorological Organization, the UNITED NATIONS ENVIRONMENTAL PROGRAMME (UNEP), and the International Council of Scientific Unions, scientists from 30 countries discussed global warming. They decided that carbon dioxide in the atmosphere was increasing mostly because of the burning of fossil fuels and the DEFORESTATION of tropical RAIN FORESTS. The plants of the tropical rain forest use much of the carbon dioxide in the atmosphere to carry out PHOTOSYNTHESIS. In this way, these plants help regulate the carbon dioxide level of the atmosphere. The scientists predicted that within the next century, Earth's atmospheric temperature could rise from 2.7° F to 8.1° F (1.4° C to 4.5° C). This would cause the levels of the oceans to rise as much as 8 to 55 inches (20 to 140 centimeters).

The scientists at the 1985 meeting also admitted that more needs to be learned about Earth's global environment. Thus, in the same year, the International Council of Scientific Unions formed the International Geosphere-Biosphere Program (IGBP). The goal of the program is to gather information that will enable scientists to understand Earth's life-support system. As part of this process, they need to study the way in which the system is being changed and the role of human activities in these changes.

The IGBP is a big project. It is also an important project. The scientists have accumulated a great deal of information. A tiny alga is now playing an important role in their research. This alga, called *Emiliania,* lives in the upper layers of the ocean. The ALGAE are abundant and are believed to make certain gases that drift upward and form thick white CLOUDS. These clouds reflect the incoming SOLAR ENERGY from the sun and help to cool the atmosphere. Studies such as this will help scientists learn whether Earth's system includes a way to counteract the greenhouse effect and global warming. [*See also* BIOGEOCHEMICAL CYCLE; CARBON; CARBON CYCLE; CHEMICAL CYCLES; METEOROLOGY; MONTREAL PROTOCOL; OZONE LAYER; SOLAR ENERGY; WEATHER; and ULTRAVIOLET RADIATION.]

Greenhouse Gas

▶ Any naturally occurring or human-made gas that contributes to the GREENHOUSE EFFECT. The major greenhouse gases are water vapor, CARBON DIOXIDE, chlorofluorocarbons (CFCS), METHANE, and NITROGEN OXIDES.

The greenhouse effect is a warming of the air that occurs when gases in the ATMOSPHERE trap heat near Earth's surface. By acting like the glass walls of a greenhouse, the gases prevent heat from escaping into space. Without the greenhouse effect, Earth would be too cold to sustain life.

INCREASING GREENHOUSE GASES

Many human activities cause the release of greenhouse gases. Carbon dioxide and nitrogen oxides are given off every time FOSSIL FUELS are burned for energy in AUTOMOBILES, power plants, and factories. CFCs enter the air during the pro-

Greenhouse Gas	Main Sources
Carbon Dioxide (CO$_2$)	burning of fossil fuels; deforestation
Chlorofluorocarbons (CFCs)	air conditioners; refrigerators; foam products; aerosol spray cans
Methane	landfills; swamps and marshes; rice paddies; livestock
Nitrogen Oxides	burning of fossil fuels

◆ Greenhouse gases are sometimes released into the air as a result of human activity. The table above lists some of the sources of greenhouse gases.

duction of plastic foam products, and from aerosol-spray cans, refrigerators, and air conditioners. Methane is released from LANDFILLS and from fossil fuels.

Many scientists are concerned that more greenhouse gases in the atmosphere could result in a warmer Earth. Studies of carbon dioxide levels in the atmosphere support this view. In the past, when carbon dioxide levels rose, the temperatures also rose. Some scientists predict that as more greenhouse gases are added to the atmosphere, GLOBAL WARMING will occur. They predict that the average temperature of Earth will increase by as much as 4° F (2° C) by the year 2050. Global warming could cause sea levels to rise and WEATHER patterns to change.

WHAT CAN BE DONE ABOUT THE CARBON DIOXIDE?

Carbon dioxide is the main greenhouse gas affected by humans. To reduce carbon dioxide emissions, environmentalists suggest that "cleaner" sources of energy could be used instead of fossil fuels. Cleaner sources include solar energy, NUCLEAR POWER, and WIND POWER, none of which release carbon dioxide.

Conservation agencies are also looking at ways to reduce the burning of tropical RAIN FORESTS, which adds billions of tons of carbon dioxide to the atmosphere each year. In addition, if more trees were planted, more carbon dioxide could be removed from the atmosphere. [*See also* AIR POLLUTION; ALTERNATIVE ENERGY SOURCES; BIOGEOCHEMICAL CYCLES; CARBON CYCLE; CLIMATE; CLIMATE CHANGE; and DEFORESTATION.]

Green Party

See GREEN POLITICS

Green Politics

▶ A term used to describe organized political activities by environmentalists. In recent years, environmentalists called *Greens* have been elected to some local, state, and national offices in western Europe. Greens support principles and practices that promote an ecologically sustainable society.

The Green movement has been strongly influenced by the peace movement, feminist movement, citizens' initiative, and other activist movements of the 1960s and 1970s. Most of the basic concerns of the Greens are similar to these groups, but the politics and strategies among different groups can vary considerably.

In the United States, environmentalism did not become a true political movement until 1962. In that year, Rachel Louise CARSON published her book called *Silent Spring*. This book helped spread awareness of issues related to the ENVIRONMENT.

By 1940, the U.S. FOREST SERVICE and such conservationist organizations as the Sierra Club, the WILDERNESS SOCIETY, the Audubon Society, and the National wildlife Federation had come into being. However, in 1962, *Silent Spring* showed Americans, as nothing had before, how human-made poisons threaten nature.

Many CONSERVATION groups benefited from the success of *Silent Spring*. For example, Sierra Club membership rose from 20,000 in 1959 to 113,000 in 1970—the year of the first EARTH DAY. Another benefit that grew from the success of *Silent Spring* was the increased willingness of members of Congress to pass laws intended to protect the environment. The first such law was the WILDERNESS ACT of 1964.

The Green movement in western Europe arose during the late 1970s and early 1980s. West German Greens won seats in the Bundestag—the national parliament —in 1983 and increased their presence four years later. Although active in national governments, Green parties stress international cooperation. They have called for making the United Nations (U.N.) and the World Court stronger and more democratic. Some have called for the abolition of the right of veto in the U.N. Security Council because only the council's five permanent members have that right. Greens from all nations have taken part in international environmental initiatives such as the UNITED NATIONS EARTH SUMMIT. The Earth Summit was the U.N. Conference on the Human Environment and Development. It was held in Rio de Janeiro, Brazil, in 1992. [*See also* AUDUBON, JOHN JAMES; FRONTIER ETHIC; GREEN REVOLUTION; MONTREAL PROTOCOL; MUIR, JOHN; and SUSTAINABLE DEVELOPMENT.]

Green Revolution

DThe name given to an international scientific effort to reduce world hunger by the use of new farming techniques and new varieties of crops in Latin American and Asian countries. The Green Revolution began in 1944, when Dr. Norman E. Borlaug and three other young American plant scientists went to Mexico to help that country's troubled agricultural program.

In 1953, Dr. Borlaug started crossbreeding varieties of wheat. He wanted to breed a new kind of wheat that was suited to conditions in tropical Mexico, where heavy rains beat down tall, slender wheat PLANTS. By 1963, he had developed a dwarf variety of wheat that grew quickly. Another improved plant was called *miracle rice*. Unlike other rice plants, it had a short, sturdy stalk that kept it from drooping and falling over. The new varieties were bred for increased harvest ratio, which is the ratio of edible to nonedible part of the plant. The Green Revolution, and Dr. Borlaug's research, is most commonly associated with the development of hybrid corn.

These and other new varieties of grain helped many countries, such as India and China, feed their ever-growing populations. India started growing high-yielding varieties in 1966. By 1978, India's total rice production had increased by as much as 20% and its total wheat

◆ Dr. Norman Borlaug (right) was awarded the Nobel Peace Prize in 1970 for his work that resulted in the Green Revolution.

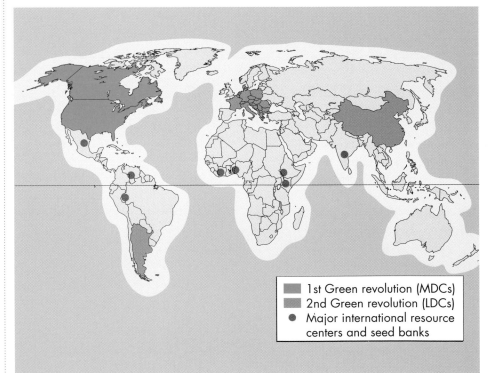

1st Green revolution (MDCs)
2nd Green revolution (LDCs)
● Major international resource centers and seed banks

◆ The Green Revolution is an international effort to aid in increasing food production in many parts of the world.

production by as much as 100%. In recognition of his outstanding contribution, Dr. Borlaug received the Nobel Peace Prize in 1970.

The Green Revolution was not a complete success, however. It did bring prosperity to parts of the world where the new varieties and methods could be used on a large scale and food could be grown for sale. But it made matters worse in other regions where its methods were not suited to the needs of SUBSISTENCE AGRICULTURE. Subsistence farmers grow only enough food to feed their families. They cannot afford to buy the expensive machinery, fertilizers, and PESTICIDES needed to grow the new varieties.

The Green Revolution's heavy dependence on machinery, fertilizers, and pesticides has also created different problems for those who can afford to use them. Machinery consumes a great deal of energy, and fertilizers and pesticides pollute the ENVIRONMENT.

Another negative result of the Green Revolution is that the old, local strains of crop plants are being replaced by the high-yielding varieties. Many scientists worry that this trend could cause a dangerous decline in the GENETIC DIVERSITY of crops. Genetic diversity is the variety of genetic differences within a single SPECIES. Without sufficient genetic diversity, a species cannot meet new challenges, and it will die out. As the potential for the use of genetically engineered crops and seeds continues to increase, so does concern about the possible associated environmental problems. [*See also* BIODIVERSITY; EXOTIC SPECIES; GENETIC ENGINEERING; HYBRIDIZATION; and SUSTAINABLE AGRICULTURE.]

Grizzly Bear

The large SPECIES of bear that was once common in much of western North America, but now occupies only a small part of its former range. Grizzly bears are one of three species of bears that live in North America. They are adapted to living in the mountains and plains of the west and the northern TUNDRA.

Large grizzlies may reach a length of 8 feet (2.4 meters) and weigh more than 1,000 pounds (450 kilograms). These bears are as fast as some horses and swim well. Although grizzlies are generally considered CARNIVORES, or meat-eaters, they actually eat a variety of foods. Their diets generally include grasses, leaves, berries, roots, grubs, eggs, and dead animals. In addition, grizzlies are PREDATORS of many INSECTS, mice, ground squirrels, lizards, snakes, FISH, deer, and other animals.

Grizzlies are not sociable animals. While they often gather at good grazing or fishing spots, they generally stay away from each other. A mother grizzly bear, who may stay with her cubs for a year or more, even has to protect her cubs from being killed and eaten by other grizzlies.

GRIZZLY BEARS AND PEOPLE

Grizzlies generally avoid people, but sometimes there are conflicts.

◆ Grizzly bears are omnivores. They eat animals such as mice, fish, and deer as well as grasses, berries, and roots.

When humans move into the bear's HABITAT, they often change the area into farmland, ranches, mines, parks, or other facilities suited to use by humans. In these new settings, grizzlies may kill LIVESTOCK or forage around GARBAGE and campgrounds, damaging property and getting too close to people. Mother grizzlies may attack humans that are near their cubs. As a result, people who settle in their habitat have tended to kill off the bears or drive them out of the area.

MANAGING GRIZZLIES

In the 1990s, some populations of grizzly bears are listed as threatened under the ENDANGERED SPECIES ACT. Outside of Alaska and western Canada in North America, their numbers have been greatly re-duced. For example, the number of grizzlies in the lower 48 United States has decreased in the last two centuries from about 100,000 to less than 1,000. Those that survive in the lower 48 states live in the mountains of Wyoming, Montana, Idaho, and Washington.

Like other ENDANGERED SPECIES, grizzlies need more than just protection from killing. It is important to find out the number of animals needed to keep the population breeding well and to find out how much habitat those animals need. Government and private organizations need to work together to provide this habitat.

Although hundreds of grizzlies still live in the American West and are now protected, ecologists suspect that these bear populations may become extinct. These ecologists believe there may not be enough protected grizzly habitat to maintain the bears' *minimum viable population (MVP)*. Minimum viable population refers to the smallest number needed to keep up a population into the future (at least for 100 years). The MVP of the grizzly bear depends on many factors, including how often it breeds, how many young it has, how long it lives, how many bears in the population breed, how much these factors differ from one bear to the next, and the likelihood that NATURAL DISASTERS will kill parts of the population. An estimated safe MVP for grizzlies is 500. A smaller number may keep the population going, but the bears remain at risk of becoming extinct.

One problem involved in preventing the extinction of the grizzly bear is that even huge pieces of land cannot support bear populations having as few as 50 members. One of the largest refuges of the grizzlies, the Yellowstone-Grand Teton National Park area, covers about 3,900 square miles (10,140 square kilometers) of land, but is still too small to support 50 grizzlies. A refuge that would provide habitat for 500 grizzlies would have to be about 12 times the area of the Yellowstone and Grand Teton National Park.

CONFLICT OVER GRIZZLIES

If NATIONAL PARKS are not large enough to keep up grizzly bear populations for long, it is important to save their habitat outside the parks, too. This presents some difficulties. The YELLOWSTONE NATIONAL PARK and Grand Teton Park, for example, are surrounded by seven NATIONAL FORESTS. The parks and FORESTS could all be managed together as a single ECOSYSTEM. However, this task is hard to organize. National parks and forests have

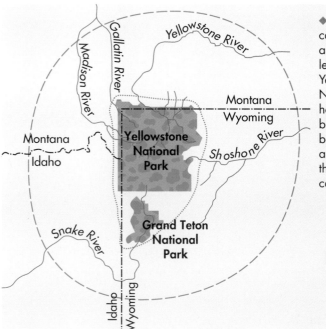

◆ Even a national park can be too small to support a grizzly population. The legal boundaries of the Yellowstone-Grand Teton National Park are shown here, together with the biotic boundaries. A biotic boundary encloses the area needed to support the smallest population that can survive.

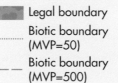

Legal boundary

.... Biotic boundary (MVP=50)

— — Biotic boundary (MVP=500)

different objectives in protecting NATURAL RESOURCES. People also disagree on the best ways to manage grizzlies. A recent federal government plan for preserving them in the lower 48 states was much criticized. Some scientists charged that the plan was based on studies that were not carefully done. Many believe the plan fails to protect habitat.

The habitat of grizzly bears outside parks is threatened partly by industries that are meeting the nation's demand for more MINERALS and oil. Large mining operations are planned for just outside the boundary of Yellowstone Park in Wyoming. Requests have been made to drill for sources of oil and NATURAL GAS near the borders of Glacier National Park in Montana. Even in Alaska, grizzly habitat is expected to shrink as the human population expands.

Because of all these factors, the fate of threatened grizzly populations is uncertain. Their survival will probably depend on how well the needs of grizzly bears can be met, while still meeting a variety of human needs. In this way, their fate is similar to that of most threatened and endangered species. [*See also* MINING; MULTIPLE USE; NATIONAL WILDLIFE REFUGE; WILDERNESS; and WILDLIFE MANAGEMENT.]

Gullying

▌▶The formation of channels in SOIL caused by running water. Running water is a common agent of EROSION. Barren, freshly plowed,

◆ This soil erosion occurred in San Miguel Island, Channel Islands National Park.

and loose SOIL, especially on sloped land, is most at risk of erosion by running water. Water that moves along furrows in plowed land or down hillsides can flow swiftly over the soil. As it moves, the water picks up and carries away soil particles. This process may be repeated over and over with each new rainfall. As soil continues to be carried away from an area, deep channels or gullies form in the land. This type of erosion is called *gullying*. Gullying is a problem for farmers because it results in a huge loss of TOPSOIL. In addition, the gullies that form make it difficult for farm machinery to move across the land.

The use of poor farming and ranching methods often help to produce gullying. For example, the clearing away of trees and other vegetation in order to prepare land for the growth of crops makes the land more vulnerable to erosion from running water. The problem is made worse if the farmland is on sloping land. Allowing cattle to overgraze their fields also places land at greater risk of gullying. [*See also* AGROECOLOGY; CONTOUR FARMING; NO-TILL AGRICULTURE; OVERGRAZING; RUNOFF; and SEDIMENT.]

◆ Serious erosion may take acres of land out of cultivation.

◆ Rainfall caused this erosion as it ran along natural depressions in the land.

Gypsy Moth

❙An exotic moth that has caused considerable damage to trees in New England and throughout the northeastern United States. The gypsy moth, *Portheria dispar,* was brought from Europe in the mid-1800s to a laboratory in Massachusetts. It was brought here for experimentation to determine if the INSECT could produce usable qualities of silk. The moth escaped from the laboratory, and, having no natural enemies in the United States, multiplied rapidly. It spread through Eastern FORESTS.

The gypsy moth lays its eggs in July, usually on tree trunks, fence rails, or fallen trees. The eggs hatch in the spring. The newly hatched furry brown, red, and blue caterpillars begin to feed on the tree leaves. Within a few weeks they strip a tree bare. Deprived of its leaves, its energy-making factories, the tree can be weakened. Several years of such depletion can kill a tree.

A large variety of techniques have been used to try to control gypsy moth outbreaks. A variety of PESTICIDES have been used. Often, these have actually resulted in even worse outbreaks when the few moths that are resistant to pesticides have produced moths that can survive some pesticide applications. Also, the pesticides tend to kill off other harmless or beneficial insects. A BACTERIA called *Bacillus thuringiensis (Bt)* has also been used with great effect, though it kills other caterpillar species as well. Sticky traps on trees and scent-lure traps have also been used. Introducing some of the gypsy moth's natural PREDATORS, such as a predatory beetle, a virus, and other enemies, has been very effective in controlling some outbreaks.

The explosion of the gypsy moth's population in New England in the absence of natural enemies is an excellent example of the impor-

tance of the predator-prey relationship to the balance of nature. Trees were being killed by the moth because it was overpopulating in the absence of a natural enemy. Gypsy moth infestations continue to be a problem at various times throughout the Northeast. They have been continuing to spread and have been found in parts of the Pacific Northwest. [*See also* BIOLOGICAL CONTROL; EXOTIC SPECIES; INTEGRATED PEST MANAGEMENT (IPM); NATIVE SPECIES; and SYMBIOSIS.]

◆ Gypsy moths feed on the leaves of an oak tree.

◆ The adult female gypsy moth (left) is larger than the male (right).

H

Habitat

▶ The place where an organism or many organisms live. Several factors make up an organism's habitat. A habitat may be thought of as the "home" of an organism. For example, the branches of a tree become the habitat of a BIRD, while the habitat of an earthworm is the SOIL. Within its habitat an organism obtains all the food, shelter, water, and other material needs essential to its survival.

TYPES OF HABITATS

Organisms inhabit virtually every available space on Earth. To survive within a particular habitat, organisms have ADAPTATIONS that enable them to carry out their life processes. For example, to survive in aquatic habitats, many FISH have a streamlined body shape and other structures that allow them to move effortlessly through the water.

Aquatic Habitats

Approximately 75% of Earth's surface is covered with water. Many SPECIES on Earth make their homes in aquatic habitats. Some aquatic habitats are marine or saltwater ENVIRONMENTS. Others are freshwater environments.

◆ Many aquatic animals such as these fish have sleek, streamlined bodies for moving efficiently through water.

OCEANS are the main marine habitat. Although oceans are vast, PLANTS and most other PRODUCERS can grow only where there is enough sunlight for PHOTOSYNTHESIS and where **nutrients** are plentiful. As a result, most ocean-dwelling organisms gather in shallow waters near the edges of continents. Here, rivers wash essential nutrients, such as calcium and phosphorus, from the land. In addition sunlight can penetrate to the bottom of shallow ocean waters, permitting the growth of ALGAE called seaweeds on the rocky bottom and PHYTOPLANKTON at the water's surface. These producers support a great variety of fish, MAMMALS, birds, REPTILES, and

INVERTEBRATES such as clams, crabs, and marine worms.

Freshwater habitats include the standing waters of lakes, ponds, and WETLANDS and the moving waters of rivers and streams. As in oceans, life is more diverse and abundant in shallow water where sunlight and nutrients are plentiful. Here, producers such as cattails, water lilies, and algae are quite abundant and attract many animals, such as frogs, turtles, fish, and INSECTS. PLANTS along the shores of lakes and ponds are rooted in the bottom mud and often possess long stems and leaves that can float on the water's surface. Freshwater animals, like other aquatic species,

◆ Zebras, wildebeests, and elephants graze on a grassland at Masai-Mara, Kenya.

◆ A squirrel eats desert grass.

◆ Capuchin monkeys live in tropical rain forests.

◆ Musk oxen live in the Northwest Territories, Canada.

◆ Douglas firs thrive in this cold forest.　　　◆ The arctic fox lives in the tundra biome.

also have adaptations for swimming through dense water.

In rapidly moving rivers and streams, plants and animals have traits that help them keep from being carried away by currents. For instance, mosses have strong root-like structures that enable them to cling to rocks. Many fish species, such as trout and SALMON, have very streamlined bodies and strong muscles to swim against currents.

Terrestrial Habitats

The land surfaces of Earth contain a variety of habitats. Land-dwelling species must adapt not only to the conditions of the particular habitats in which they live—soil, air, trees—but also to the CLIMATE of the area.

Climate varies greatly around the world. However, all terrestrial species adapt to their habitats in remarkably similar ways.

THE LANGUAGE OF THE ENVIRONMENT

burrows underground tunnels or holes made by animals.

nutrients substances such as food, water, oxygen, and nitrogen that are critical for life.

Arboreal, or tree, habitats are one of the most dominant habitats in FORESTS. Almost all animal groups and some plant groups have arboreal species. Arboreal species have many adaptations for survival in trees. For instance, small animals, such as squirrels, tree frogs, and lizards, have claws or friction pads on their fingers and toes to help them grip tightly the bark or branches of a tree. Larger arboreal animals, such as monkeys, apes, and sloths, have powerful arms suited to climbing.

Many plants, such as tropical orchid species, survive in dense forests by growing high above the ground in the branches of trees. There, the plants obtain the sunlight they need for photosynthesis. Water is taken directly from the moist, tropical air.

GRASSLAND habitats are characterized by open, grassy areas that contain small plants and a few scattered trees. As such, open ground is the main habitat in both temperate grasslands and tropical SAVANNAS.

GRAZING species, such as bison, are common in grasslands. When those species eat all the grasses in one area, they migrate to new areas to find freshly sprouted grasses. Such species sometimes travel long distances to find new grass, so most grazers are adapted for moving quickly.

Other grassland animals, such as prairie dogs, badgers, insects, and some birds, build extensive underground **burrows** to shield them from WEATHER conditions and protect them from PREDATORS. Plants adapted to grassland habitats often have large underground root systems that can survive fires and droughts during scorching dry seasons.

DESERT habitats are very dry. Plants and animals that live in deserts have adaptations for storing and conserving water. Most desert plants have thick stems and leaves that store water. Some have waxy coverings on their stems and leaves to help to reduce water loss. Desert animals often protect themselves from the high temperatures by burrowing into the soil where it is cooler during daytime hours.

TUNDRA habitats are characterized by extremely cold temperatures and dry conditions. Most of the water in arctic and tundra regions is locked away in ice and in *permafrost*—permanently frozen soil. Vegetation in these types of habitats is small and sparse. Plants that do exist typically are short and hug the ground for warmth. They are also adapted to take advantage

of the limited sunlight of the short growing season. They grow and flower quickly.

Animals living in tundra regions are also well adapted for the cold temperatures. MAMMALS typically have dense, shaggy coats and layers of fat to protect and insulate them from the cold. Many tundra animals also have coloring that changes at different times of the year to help them blend in with their surroundings to avoid predators. [*See also* BIOME; ECOSYSTEM; and HABITAT LOSS.]

Habitat Loss

▶The removal or destruction of HABITAT where an organism lives. In its habitat, an organism obtains food, water, sunlight, MINERALS, and other substances that are critical to its survival.

Habitat loss is a significant threat to Earth's BIODIVERSITY. It is difficult to estimate how many SPECIES become extinct each year because of habitat loss. However, most scientists and world conservation groups agree that habitat loss is the single greatest cause of EXTINCTION today. A 1994 study by the World Conservation Monitoring Center determined that 36% of all animal extinctions are caused by habitat destruction. Similarly, biologist Edward O. WILSON of Harvard University estimates that about one-third of all extinctions are the result of habitat loss. Other major causes of extinction include the introduction of EXOTIC SPECIES into an area, HUNTING, the use of various PEST CONTROL methods, and POLLUTION.

HABITAT LOSS AND HUMAN ACTIVITIES

Most habitat loss is caused by the growth and expansion of the human population. As the population grows, more land is needed for housing, farms, roads, cities, bridges, and DAMS. Land and water habitats are also destroyed by

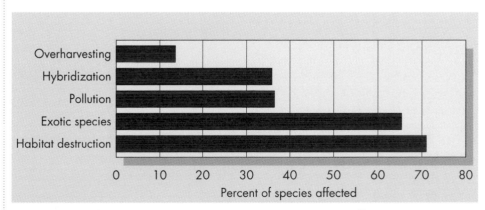

◆ Among the different threats to biodiversity, habitat destruction affects the greatest number of species.

 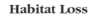

increased human use of NATURAL RESOURCES, such as FUEL WOOD, timber, water, FOSSIL FUELS, minerals, and metals. When humans use land and water for these purposes, habitats can be destroyed. Because species are adapted to live under certain specific conditions only, loss of even one part of an organism's habitat can lead to extinction.

Today, oceans, lakes, and rivers can be damaged and destroyed by pollution. Grasslands can be changed into DESERT through such activities as OVERGRAZING by LIVESTOCK and SALINIZATION of the SOIL caused by IRRIGATION. However, DEFORESTATION of the tropical RAIN FORESTS of Africa, Southeast Asia, and South America represents the greatest threat to biodiversity today. Deforestation is the clearing of a forest for use of its products or land. The removal of timber; use of land for agriculture, such as cattle ranching and the growth of crops; and large-scale building of roads and DAMS are common reasons for deforestation.

Tropical rain forests cover only about one-sixth of Earth's surface, but they contain close to one-half of all known species. For instance, a 2,500-acre (1,000-hectare) region of tropical rain forest contains as many as 1,500 species of FLOWERING PLANTS, 750 species of trees, 400 species of BIRDS, and countless INSECTS. These numbers do not include the thousands of species of microscopic organisms such as BACTERIA and protists, or the many species of FUNGI, that lie in the soil, air, water, or even in other organisms. Wilson has estimated that 4,000 to 6,000 species are being eliminated each year due to rain forest destruction.

◆ Logging activities in the rain forest of Papua, New Guinea, may result in the loss of diversity in the area.

◆ The clear-cutting of tropical rain forests worldwide destroys the habitats of countless organisms.

PROTECTING HABITATS

Experts agree that the best way to preserve Earth's biodiversity is to protect habitats from being destroyed. To achieve this goal, the United States has set aside tens of millions of acres of WILDERNESS lands that are protected from all kinds of development. However, decisions on how land should be used can sometimes be controversial. Such a situation arose in 1990 when the NORTHERN SPOTTED OWL became the center of debate on how OLD-GROWTH FORESTS located in the northwestern United States should be used.

When the U.S. FISH AND WILDLIFE SERVICE listed the northern spotted owl as a threatened species, millions of acres of forest in Washington and Oregon were designated as protected habitat. Under the ENDANGERED SPECIES ACT, no trees could be cut down within the owls' protected forest. Loggers, sawmill workers, and residents of logging communities complained about this decision. They did not think it was fair that their jobs were being threatened to save a single bird species from extinction. In 1995, the Supreme Court ruled that the laws should be upheld and the northern spotted owls' habitat protected. However, the controversy over this protected habitat is not likely to end soon. [*See also* ABIOTIC FACTORS; AIR POLLUTION; BIOME; BIOREGION; CLEAR-CUTTING; CORAL REEF; DEBT FOR NATURE SWAPS; DEPARTMENT OF THE INTERIOR; ENDANGERED SPECIES; LAND USE; MARINE POLLUTION; MULTIPLE USES; NATIONAL GRASSLAND; NATIONAL PARKS; NATIONAL SEASHORE; NATIONAL WILDLIFE REFUGE; PUBLIC LAND; WATER POLLUTION; and WILDERNESS ACT.]

◆ The Hanford Nuclear Reservation has more than 1,000 waste sites that must be cleaned up.

Hanford, Washington

▶ The site of a U.S. government reservation, where nuclear materials were made during and after World War II. Hanford was a small farming town located along the Columbia River in eastern Washington State.

During World War II, the U.S. government took over the 300-mile (480-kilometer) area around Hanford to build nuclear reactors and create other facilities to make materials for NUCLEAR WEAPONS. Hanford was an ideal location. It was remote, but located near a ready source of water for cooling nuclear reactors.

People from the Hanford area and nearby White Bluffs were moved out. Hanford grew into a major research and production facility, now overseen by the Department of Energy. Wastes from the NUCLEAR POWER plants and pro-

cessing sites were managed in very different ways. Today, the Hanford nuclear reservation is known to have over 1,000 waste sites that need to be cleaned up. Under the SUPERFUND laws, the cleanup will take place over the next 25-50 years at a cost of up to $250 billion. [*See also* COMPREHENSIVE ENVIRONMENTAL RESPONSE, COMPENSATION AND LIABILITY ACT (CERCLA); HAZARDOUS SUBSTANCES ACT; HAZARDOUS WASTE; HAZARDOUS WASTE MANAGEMENT; RADIOACTIVE WASTE; TOXIC WASTE; and WATER POLLUTION.]

Hardin, Garrett (1915–)

▶ Ecologist and educator credited with exploring some issues related to the moral and social effects of science on society. Hardin once

studied microorganisms and drama. In over 30 years as Professor of Human Ecology at the University of California at Santa Barbara, Hardin has developed a reputation as a person who does not flinch from the task of telling unpleasant truths.

Hardin is best known for his 1968 essay "The Tragedy of the Commons," published in *Science* on December 13. The essay shows how people using a common resource, such as a "common" used for grazing cows, tend to overload the common for personal gain at the expense of others. In other words, a herdsman who adds one too many cows to the common diminishes the gain of all the users, but gains by the addition of a cow to his own herd. Hardin compared this overloading of the common ground to the growth of human populations, which overloads the natural landscape. He argued that the uncontrolled freedom to breed was as unethical as the herdsman taking more than a fair share.

While Hardin has focused much of his writing on limiting human POPULATION GROWTH, his work has had a strong influence on the regulation of a variety of environmental problems. Hardin has continued to argue for and write about population control while cautiously revealing many social taboos that he believes contradict science and common sense. [*See also* ENVIRONMENTAL ETHICS; SUSTAINABLE DEVELOPMENT; and TRAGEDY OF THE COMMONS.]

Hazardous Materials Transportation Act (HMTA)

▶A law designed to prevent the illegal dumping of hazardous substances on highways, railways, and vacant lots. A hazardous substance poses a significant threat to human health or the ENVIRONMENT.

The main producers of hazardous substances are chemical and PETROCHEMICAL plants. Hazardous substances are also contained in many of the products people use every day, such as the chemical PESTICIDES used on farms, lawns, and gardens.

PASSING HMTA

The Hazardous Materials Transportation Act (HMTA) was passed in 1975 to replace the Hazardous Materials Control Act of 1970. The earlier act was ineffective and not properly enforced. During the 1970s, illegal dumping of hazardous materials increased partly due to greater production of synthetic chemicals. The problem was made worse because many LANDFILLS refused to take HAZARDOUS WASTES. Rather than spending money and time to dispose of the materials properly, some individuals illegally dumped them in remote areas.

TRANSPORTING HAZARDOUS SUBSTANCES

HMTA establishes a list of minimum standards for the transport of hazardous substances by airplanes, ships, trains, and motor vehicles. Under these laws, all hazardous waste transporters must use a special tracking system to keep careful records of the pickup and delivery of all hazardous wastes. Wastes can be delivered only to recognized hazardous waste sites, and proper agencies must be notified immediately if any accidents occur during the transportation process. In addition, all vehicles carrying hazardous wastes must display special stickers indicating the types of hazardous wastes on board. The U.S. Department of Transportation is responsible for enforcing the laws of HMTA.

UPDATING HMTA

In 1990, the Hazardous Materials Transportation Uniform Safety Act was passed. These laws strengthened those already included under HMTA. Changes included increasing the number of safety inspectors, increasing the criminal penalties for violations of the laws, and helping states develop better response

♦ Under the Hazardous Materials Transportation Act, all vehicles carrying hazardous materials must display stickers to notify the public about the types of materials being transported.

plans for hazardous waste accidents. [*See also* CANCER; CARCINOGEN; HAZARDOUS SUBSTANCES ACT; HEALTH AND DISEASE; HEAVY METALS POISONING; HERBICIDE; INSECTICIDE; LEACHING; RESOURCE CONSERVATION AND RECOVERY ACT (RCRA); RODENTICIDE; TOXIC SUBSTANCES CONTROL ACT; and TOXIC WASTE.]

Hazardous Substances Act

▌A law passed in 1960, officially named the Federal Hazardous Substances Act, which bans the use of certain hazardous substances and sets rules for labeling household products that contain hazardous substances. Many products that are used around the home contain hazardous substances. The substance may be toxic, may cause possible damage to the skin or eyes, or may cause ill effects from its fumes if the product is not used carefully. The Federal Hazardous Substances Act was passed so that people would be warned of the possible dangers of the products they use.

The act requires that a warning be printed on the front label of any product that contains a hazardous substance. The warning must describe the **hazard,** telling whether, for example, the product is harmful if swallowed, spilled on the skin, or inhaled. The label must also include directions for using the product safely. These might tell the user of the product to wear gloves, to use the product only where there is plenty of fresh air, or keep the product away from food, pets, or plants. The manufacturer of the product must make sure this information is correct. Putting the wrong information on a warning label is considered a crime.

The Hazardous Substances Act is only one of many laws that control the handling of hazardous substances. Since this act is meant to protect consumers, it is overseen by the Consumer Products Safety Commission (except in the case of products already covered by the ENVIRONMENTAL PROTECTION AGENCY (EPA) and the Food and Drug Administration (FDA). The act does not require a label to give directions for safely disposing of a product. It is often up to the user to find out if a product can be dumped down a drain or put into a trash can, or if it should be treated as TOXIC WASTE and sent to a **disposal**

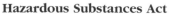

THE LANGUAGE OF TH**E** ENVIRONMENT

disposal center place in a community to bring hazardous household wastes (such as leftover paint thinner, pesticides, old car batteries, and so on) for proper disposal. The center helps keep hazardous substances out of local sewage systems and landfills. Some communities also have collection days for hazardous wastes.

hazard source of danger.

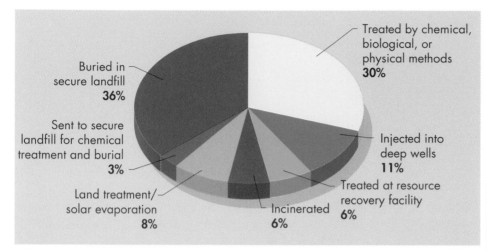

◆ The most common disposal practice for hazardous waste is burial in secure landfills.

Buried in secure landfill **36%**

Treated by chemical, biological, or physical methods **30%**

Sent to secure landfill for chemical treatment and burial **3%**

Injected into deep wells **11%**

Land treatment/ solar evaporation **8%**

Incinerated **6%**

Treated at resource recovery facility **6%**

center instead. [*See also* CARCINO-GEN; FEDERAL INSECTICIDE, FUNGICIDE, and RODENTICIDE ACT (FIFRA); HAZARDOUS WASTE; HEALTH AND DISEASE; LANDFILL; SOLID WASTE; and TOXIC WASTE.]

MENTAL PROTECTION AGENCY (EPA) as hazardous because they are inflammable, corrosive, unstable, or toxic. Many people argue that cancer-causing and RADIOACTIVE WASTE should also be included in the EPA's list of hazardous materials.

Disposing safely of hazardous waste is a major problem. The waste is usually sealed in containers and buried in special LANDFILLS that are monitored for leakage and expensive to use. Some waste is buried in deep wells or in caverns. The danger is that these underground dumps will eventually leak or be damaged by earthquakes and release TOXIC WASTE into the groundwater.

Leaking containers caused LOVE CANAL, New York, to become the first waste site declared a federal emergency disaster area. In the 1950s, Hooker Chemical Company sealed hazardous waste into metal drums and buried them in a ditch, which was covered with clay. Later, homes and other buildings were constructed on the land. By 1977, toxic wastes were leaking from the ditch. About 1,000 families were

Hazardous Waste

◗ Waste that is flammable, capable of releasing toxic fumes, or from which harmful concentrations of toxic materials can leach. Each year, the United States produces about 275 million tons (250 metric tons) of hazardous waste. This is more than any other country. Even our household waste includes PESTICIDES, cleaning products, and furniture polish, all of which are toxic. You sometimes see trucks or railroad cars labeled "Hazmat" (Hazardous Material). These contain substances defined by the ENVIRON-

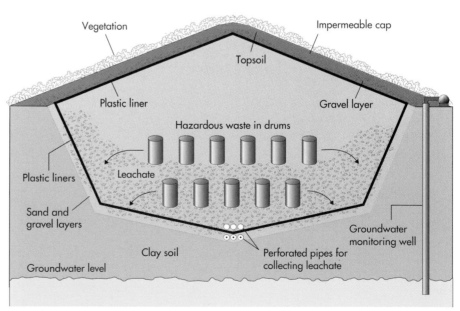

◆ To make a hazardous waste landfill as secure as possible, plastic liners and layers of sand and gravel are placed underneath. Various systems may be built in for monitoring and controlling leachate and runoff.

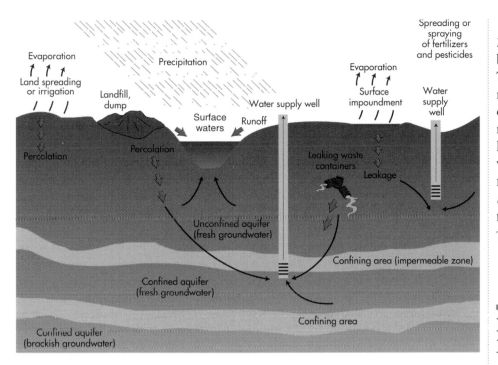

◆ In time, hazardous wastes disposed of on land may leak out of their containers and contaminate the aquifers.

In 1992, producers saved about 27 million dollars in disposal fees by exchanging hazardous wastes. The U.S. Department of Energy sold radioactive cesium to a Canadian company that used it to make medical products. The cesium was hazardous waste from HANFORD, WASHINGTON, once a major site for NUCLEAR WEAPONS production. [*See also* HAZARDOUS WASTE MANAGEMENT; RECYCLING, REDUCING, REUSING; and TOXIC SUBSTANCES CONTROL ACT.]

Hazardous Waste Management

See TOXIC WASTE and TOXIC SUBSTANCES CONTROL ACT

Hazardous Wastes, Storage and Transportation of

▶ Methods of securing toxic, radioactive, or otherwise dangerous waste materials in one place and of moving them from place to place. A wide variety of substances are labeled HAZARDOUS WASTES, including toxins in household trash, industrial waste, and high-level RADIOACTIVE WASTE. Some wastes can be made less dangerous by using special methods of heating or burning. Others are made safer through BIOREMEDIATION or treatment with chemicals.

relocated after they developed high rates of birth defects, miscarriages, liver CANCER, and nervous disorders. By 1990, part of Love Canal had been cleaned up. However, 1,200 other abandoned hazardous waste sites, known as SUPERFUND sites, have been identified.

Some toxic substances can be easily converted into substances that are not hazardous. For instance, combining cyanides with OXYGEN forms CARBON DIOXIDE and nitrogen. Surplus stores of the HERBICIDE Agent Orange were disposed of by burning them. In 1993, researchers isolated BACTERIA that break down toxic polychlorinated biphenyls (PCBS) into carbon dioxide and water.

The most effective solution to the problem of hazardous waste is to produce less of it. The chemical industry encourages changes in manufacturing methods in order to produce less waste. Today, books are often printed with soy-based inks and held together with BIODEGRADABLE glue, thereby reducing the amount of hazardous waste produced by the printing industry.

Hazardous waste may sometimes be reused instead of being discarded. Several programs help companies that produce waste find other companies that can use the waste. One list of waste items from the Pacific Materials Exchange included 10,000 pounds (4,500 kilograms) of mink fat, 100 pounds (45 kilograms) of gold glitter, and 800 gallons (3,000 liters) of cyanide solution. Among the wanted items were fish waste, 100 gallons (375 liters) of used antifreeze, and sour milk.

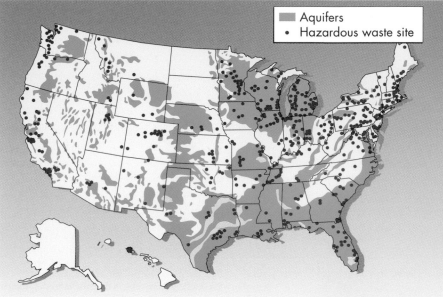

Aquifers
• Hazardous waste site

◆ Many hazardous waste disposal sites are near aquifers. If wastes break out of their containers, they could pollute the nearby drinking water.

◆ Transporting hazardous wastes can be dangerous. One way to avoid moving hazardous wastes is to use a mobile incinerator, which can burn the wastes where they are produced.

Not all hazardous wastes can be made less dangerous. Radioactive wastes, in particular, are difficult to make safer because they often remain radioactive for many years. To help reduce the harmful effects of hazardous wastes, it is necessary to store them permanently in places where they cannot leak into the SOIL, air, or water.

METHODS OF STORAGE

Many hazardous wastes are stored in secured LANDFILLS. These are pits in the ground that have been constructed with thick clay bottoms and tough linings to help prevent the wastes in the landfill from leaking into the soil or groundwater. Other storage places for hazardous wastes include deep wells, caverns, underground storage tanks, warehouses, SLUDGE pits, and lined storage ponds.

In theory, storage sites for hazardous wastes prevent the wastes from escaping into the ENVIRONMENT. However, almost any tank, landfill, or other container will develop a leak sooner or later. Thus, it is important to check the environment around a storage site regularly for signs of escaping waste. The 1984 Hazardous and Solid Waste Amendments require that the groundwater near landfill sites be checked for a period of 30 years after the site is closed. However, no precautions have been developed for containing and monitoring hazardous wastes in landfills closed for more than 30 years.

Storing radioactive waste is a special problem. In the United

States, plans for storing radioactive wastes permanently in large caverns (deep underground caves) dug out of rock or salt have been developed. Every storage site must be studied carefully before it is used because the cavern must be dry and safe from earthquakes and other disruptions. Suitable sites are difficult to locate. A fairly active volcano was discovered near one possible storage location in Yucca, Nevada. Groundwater leaks were found at another site near Carlsbad, New Mexico. While the search for permanent storage places goes on, radioactive wastes are currently kept in storage areas above ground.

TRANSPORTING HAZARDOUS WASTE

At one time, producers of hazardous wastes could get rid of them without reporting where they went. This has changed, in part because of the RESOURCE CONSERVATION AND RECOVERY ACT (RCRA). Now, anyone who handles hazardous materials must file documents that account for the location of such materials. In this way, the materials are traced from their development through their final storage or disposal. This practice for transporting hazardous wastes has reduced POLLUTION, but many hazardous wastes are still released into the environment through leaking storage sites and illegal dumping. [See also COMPREHENSIVE ENVIRONMENTAL RESPONSE, COMPENSATION, AND LIABILITY ACT (CERCLA); INDUSTRIAL WASTE TREATMENT; LEACHING; NIMBY (NOT IN MY BACKYARD); SUPERFUND; TOXIC WASTE; TOXIC WASTE, INTERNATIONAL TRADE IN; and WASTE REDUCTION.]

Health and Disease

▶ States of the functioning of an organism. Health and disease are sometimes considered as opposite states, with health representing normal functioning of an organism and disease representing an abnormal condition. By analogy, the concepts of health and disease have been applied to human societies and to ECOSYSTEMS.

DEFINING HEALTH AND DISEASE

Health is a difficult concept to define. In one sense, health refers to a state when the body functions at its best or optimum. We know that each person is not necessarily

functioning at his or her best, so health is sometimes defined as a normal range of functioning. For example, a highly trained athlete might be functioning "optimally," but many other people would be considered "normal" and healthy even though they are not trained athletes. Health can also be thought of as the absence of disease, but, again, simply being without disease does not make a person healthy.

Disease is a state in which functioning is not normal. A number of factors produce diseases in people. The causes of disease range from biological factors to environmental factors. Diseases that last a short time are called *acute* conditions; those that last a long time are called *chronic* conditions. The causes of disease may come from outside the organism or from internal, inherited

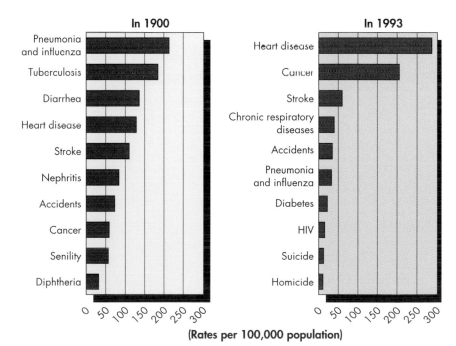

♦ In the United States at the turn of the century, pneumonia, influenza, and tuberculosis were the leading killers. Today, heart disease and cancer are responsible for the greatest number of deaths.

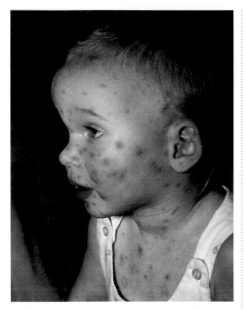

◆ Chicken pox is caused by one of the herpes-type viruses. The symptoms are fever, headache, and skin rash.

factors. Examples of outside factors include VIRUSES, BACTERIA, and parasites that cause diseases like measles, smallpox, food poisoning, or malaria. In these diseases, other organisms called PATHOGENS, invade the body and cause the body to function abnormally. The length of the disease usually depends on the ability of the body to fight it, usually through the immune system, so that many diseases, such as the common cold or measles, caused by viruses and bacteria disappear fairly quickly. However, other diseases, such as polio or AIDS, may last much longer. The length and severity of a disease also depends on the ability of medical doctors to treat it with drugs. For example, before the development of antibiotics such as penicillin, infections killed more people than chronic diseases. Now more people die from chronic diseases like heart dis-

ease than from infectious diseases. Other outside factors that produce disease include habits like smoking and environmental factors such as chemical POLLUTION.

Some diseases are actually genetic defects that may be inherited in families. For example, people with hemophilia, an inherited disease, have blood that does not clot easily. Similarly, the chance of a person developing diseases such as diabetes, heart disease, and some kinds of CANCER is affected by GENETICS.

ENVIRONMENTAL HEALTH

Environmental health scientists study the effects of environmental hazards on people. Often the goal of such studies is to use RISK ASSESSMENT to determine the risk to people of certain environmental factors that might cause problems or disease. In other words, the term *environmental health* usually refers to the effects that the ENVIRONMENT

might have on people, not the effects people have on the environment. Today's human environment presents people with a number of hazards—biological, chemical, physical, and cultural—that might produce health problems.

Biological hazards can cause diseases that people may get from the way the environment is managed. For example, improper treatment of SEWAGE can produce waterborne diseases. The diseases are spread when people drink water contaminated with viruses, bacteria, protozoans, or other parasites.

Chemical hazards can also produce disease. People can be poisoned by harmful chemicals in the air, water, SOIL, or food. For example, a HEAVY METALS POISONING produced MINAMATA DISEASE in people who ate food contaminated with MERCURY. People have been harmed by polychlorinated biphenyls (PCBS) and other chemicals improperly released into the environment, including PESTICIDES.

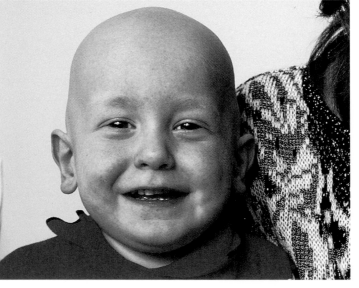

◆ When a cancer patient undergoes chemotherapy, the hair may fall out and then grow back at a later time.

Physical hazards that arise from environmental management problems include RADIATION, noise, and even flooding caused by DEFORESTATION or improperly built DAMS that collapse. Another hazard is caused by RADON. In some places, people exposed to high levels of radon increase their risk of developing lung cancer.

Cultural hazards are produced when people create unsafe working or living conditions or when they develop unsafe habits like smoking or taking drugs. Tobacco smoke, for example, is considered to be one of the causes of indoor AIR POLLUTION, as well as lung diseases, and may cause disease even in people who do not smoke.

ECOSYSTEM HEALTH

Ecologists and environmental scientists have recently drawn a parallel between environmental health, factors in the environment that affect people, and the health of natural ECOSYSTEMS. That is, a "healthy" ecosystem can be considered to be one that is operating normally. For ecosystems, this usually means that NATIVE SPECIES are present in sufficient numbers and with sufficient strength to resist most natural and some human-caused disturbances. Some scientists believe that a healthy ecosystem could better resist a pollution accident.

If ecosystems can be healthy, then there must be ecosystem "diseases" and, perhaps, ecosystem medicine. Scientists are now trying to determine what kinds of problems could be considered ecosystem diseases, what the causes are, and what the treatments might be.

Environmental scientists would like to know if there are general patterns of ecosystem disease (called *ecosystem distress syndromes*). Some of the treatments for ecosystem disease might come from the new field of RESTORATION BIOLOGY. [*See also* ASBESTOS; BIOACCUMULATION; CARBON MONOXIDE; CHLORINATION; DIOXIN; FAMINE; HEALTH AND NUTRITION; HYPOXIA; INFANT MORTALITY; MALNUTRITION; RADIATION EXPOSURE; RISK ASSESSMENT; SAFE DRINKING WATER ACT; SEWAGE TREATMENT PLANT; SICK BUILDING SYNDROME; WATER, DRINKING; and WATER QUALITY STANDARDS.]

Health and Nutrition

▶ Two words usually linked together because both refer to the body of a living thing: *health* refers to the ordinary disease-free state of the body, *nutrition* to the process by which the body uses food for growth. Someone who is healthy follows basic rules, commonly learned at a young age. The rules include brushing and flossing teeth, getting adequate rest, exercising, having yearly medical and dental checkups, washing and bathing regularly, and eating balanced meals—those with the proper amounts of essential nutrients.

NUTRIENTS AND THEIR FUNCTIONS

All living things need a source of energy and a steady supply of CARBON and nitrogen. PLANTS get their energy from the sun, carbon from the air, nitrogen from the soil, and can make their own food. Animals, including humans, must acquire their energy, carbon, and nitrogen from the nutrients in foods they eat.

Protein is essential for life; it is in every cell of the body. A constant supply is necessary to provide energy and to build and repair cells.

Proteins are complex substances that consist of smaller parts called *amino acids*. There are about 22 amino acids, and all but 8 of them can be produced by the body itself, like the **keratin** in hair and nails. But the other 8 essential amino acids must be supplied by foods. Milk, cheese, eggs, fish, and meat are complete proteins—they contain all essential amino acids. LEGUMES, nuts, cereal grains, and vegetables are incomplete proteins—they lack one or more essential amino acids. Two incomplete proteins, like peas and rice, mixed and eaten together can create complete proteins.

A lack of protein can cause loss of energy, lower resistance to disease, and **edema**. Over time, the body takes needed protein from the liver and muscle tissues, causing damage to them.

Carbohydrates are food sources found as sugars, starches, and **fiber**. Simple sugars are the carbohydrates in things like candy and soft drinks. Complex carbohydrates are found in things like pasta, potatoes, beans, and whole-grain breads and cereals. Nutritionists say 60% of a person's daily **calories** should come from carbohydrates. Complex carbohydrates furnish more vitamins and minerals than simple

calories the units for measuring heat or energy released by food digested in the body.

cholesterol the fatty substance produced in the human liver and also found in foods like eggs, meat, and dairy products. Small amounts of cholesterol help the body; large amounts can lead to heart disease.

edema a disease in which the body swells because of excessive liquid retained in the body tissues.

fiber the "roughage" from the tough cell walls of food plants that helps keep food moving through the digestive system.

keratin the tough, fibrous protein substance that forms the outer layer of hair, nails, horns, and hooves.

products. People need some fat to fuel their bodies and to promote growth. But too much can clog arteries around the heart, leading to heart attacks. Saturated fats in meat, dairy products, and coconut and palm oils create a greater risk for heart disease than do the unsaturated fats in safflower, sunflower, corn, olive, canola, and soybean oils, and some fish. Nutritionists recommend eating low-fat or fat-free foods whenever possible.

Vitamins are organic compounds necessary to regulate body functions. Some, like Vitamin C, dissolve in water, so the body needs more to replace the loss.

Others, like vitamins A and D, are stored in the liver.

Each vitamin has a specific purpose. For example, Vitamin A helps maintain living tissue and promotes bone growth, and a deficiency may cause eye problems, lung infections, and anemia. Vitamin C helps bones form and the body heal; a deficiency may cause bone deterioration and failure of wounds to heal. Vitamin D helps teeth and bones use calcium for growth; a deficiency leads to tooth decay and bone deformities.

Sources of Vitamin A are fruits, carrots, pumpkin, spinach, sweet potato, winter squash, and cauli-

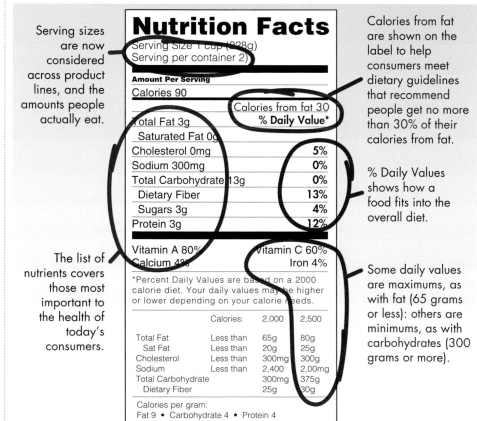

sugars. They also provide the important dietary fiber that simple sugars do not have.

Fats and oils (liquid fats) are sources of stored energy found in most foods—especially in butter, margarine, cooking oils, and animal

◆ Ingredients must be listed on Nutrition Facts labels in order by weight, making it easier for consumers to pick foods that contain the most healthful nutrients.

Important Minerals and Vitamins and Their Sources		
Minerals	**Food sources**	**Functions in the body**
Sodium	Table salt, meat, processed foods	Helps strengthen transmission of signals in the muscles and nerves
Potassium	Whole grains, fruits, meat, legumes	Regulates nerve and muscle action and helps in protein synthesis
Vitamins	**Food sources**	**Deficiency symptoms**
Vitamin A (Rotinol)	Green and yellow vegetables, butter, liver, dairy products, fish oils, sweet potatoes	Night blindness; dry, flaky skin; dry mucous membranes
Vitamin B_1 (Thiamine)	Fruits, milk, vegetables, yeast, whole grains and enriched bread and cereals, nuts, peas, lean pork	Beriberi, fatigue, heart failure, edema
Vitamin B_2 (Riboflavin)	Milk, cheese, eggs, fish, liver, poultry, vegetables	Sores on lips, bloodshot eyes, sensitivity to light
Vitamin B_3 (Niacin)	Whole grains, liver, fish, yeast	Fatigue, skin eruptions, digestive disturbances
Vitamin C (Ascorbic acid)	Citrus fruits, strawberries, brussels sprouts, cabbage, cauliflower, tomatoes, cantaloupe	Scurvy, bone deterioration, slow healing
Vitamin D (Calciferol)	Fish oils, egg yolk, milk and dairy products	Rickets

flower. Vitamin C is in citrus fruits, strawberries, brussels sprouts, cabbage, cauliflower, and tomatoes. Vitamin D is found in milk, eggs, and liver. Nutritionists have determined that humans also need 11 other vitamins in their diets: choline, niacin, pantothenic acid, Vitamin B_1, Vitamin B_6, Vitamin B_9, Vitamin B_{12}, folic acid, biotin, Vitamin E, and Vitamin K.

MINERALS are nonorganic elements important for helping the body work properly. For example, calcium builds strong teeth and bones and helps blood clot and muscles contract. Iron builds and maintains healthy red blood cells, as does COPPER. Phosphorus builds bones and is part of deoxyribonucleic acid (DNA), as are zinc and iron. And potassium is critical for a normal heartbeat.

Sources of calcium are milk, yogurt, cheese, fresh SALMON, collard greens, and broccoli. Sources of iron are lean beef, liver, whole-grain bread and cereal, spinach, and strawberries. Potassium is found in bananas, tomatoes, orange juice, and potatoes.

Other important mineral nutrients include sulfur, sodium, chlorine, magnesium, iodine, fluorine, and chromium. Sodium and chloride together make salt, which is important to regulate the body's fluid content. However, too much salt can lead to high blood pressure. Fluorine is found in fluoridated drinking water; magnesium in chocolate, seafood, and cereals; sulfur in protein foods; iodine in salt; and chromium in yeast and organ meats like liver.

WATER

Living things are made up mostly of water—humans are about 65% water. Doctors say most people should drink at least eight glasses of water a day to help regulate their bodies. Water aids digestion, carries other nutrients throughout the body, stabilizes body temperature, and removes body wastes.

HUMAN NUTRITIONAL NEEDS

To help Americans plan balanced meals, in April of 1992 the U.S.

DEPARTMENT OF AGRICULTURE (USDA) made public its Food Pyramid, which is made up of six sections of varied sizes. Each section represents one food group and lists the number of daily servings from that group that a healthy person over age two should have. (Children under age two have special nutritional needs.) Young people normally need at least the minimum number of servings from each food group to stay healthy.

The large bottom section, or foundation, includes foods that provide a good foundation for a daily diet—bread, cereal, rice, and pasta. These foods are rich in complex carbohydrates, dietary fiber, and important vitamins and minerals, and people should choose the most servings (6–11 per day) from this group. The building blocks of nutrition come next: vegetables (3–5 servings); fruit (2–4); milk, yogurt, and cheese (2–3); meat, poultry, fish, dry beans, eggs, and nuts (2–3). The top of the pyramid includes things high in fat, **cholesterol**, sugar, and/or sodium from which people should choose the

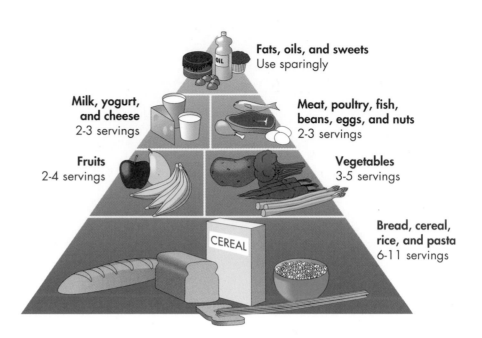

◆ The USDA Food Pyramid makes it easier for people to choose foods that add up to a balanced diet.

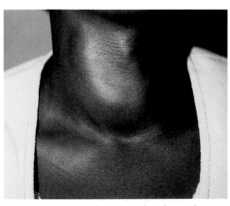

◆ A goiter is produced in the thyroid gland when there is too little iodine in the diet.

fewest servings—fatty foods, oils, and sweets, which should be eaten sparingly.

Calories measure how much energy is released by digested food. How many calories a person needs depends on his or her age, sex, height, weight, and metabolic rate—the rate at which the body burns calories. Young people, on average, need from 2,000 to 2,500 calories a day.

To help consumers count calories, the Food and Drug Administration (FDA) requires manufacturers to put Nutrition Facts labels on their products. The label must list each ingredient by its most common name and in descending order by weight. The labels also list the amounts of sugar, fats, protein, cholesterol, vitamins A, C, and D, and other nutrients in the product, and what percent of the daily value (PDV) for a nutrient is supplied by

a measured serving of the food product.

HEALTH, NUTRITION, AND THE ENVIRONMENT

The food and water that people consume come from the ENVIRONMENT, provided either by nature or through the work of farmers and other producers. Agencies of the U.S. government and environmentalists monitor agricultural and water resources to protect people from drinking water or eating foods contaminated by harmful chemicals. Consumers can help by recycling, reducing, and reusing garbage and trash to protect the environment that provides their nutritional sources. [*See also* AGROECOLOGY; CLEAN WATER ACT; FAMINE; HEALTH AND DISEASE; INFANT MORTALITY; MALNUTRITION; and SAFE DRINKING WATER ACT.]

Heavy Metals Poisoning

▌The ill effects that result from exposure to any of several metallic elements. A number of elements are called heavy metals. They are generally toxic, but how toxic they are depends on the type of metal, the amount, and what other elements it is combined with. Heavy metals occur naturally in Earth's crust, but most harm people only when heavily used for technology or industry. Processes such as MINING, manufacturing, or the burning of FUELS can release heavy metals into the ENVIRONMENT in unusually large amounts. This can damage ECOSYSTEMS or directly affect human health. Among the heavy metals that cause the most concern are arsenic, cadmium, LEAD, MERCURY, and selenium.

Heavy metals poisoning occurs when people are exposed to these elements in the workplace or elsewhere in the environment. Depending on the metal and the dose, a heavy metal can cause anything from mild illness to crippling disease, insanity, or death. Some forms of heavy metals poisoning also cause CANCER and birth defects. The long-term effects of some heavy metals as pollutants in the environment are not well known.

ARSENIC

Very small amounts of arsenic are a normal part of SOIL, water, SEDIMENTS, and all living things. People use arsenic to make metal products, batteries, PESTICIDES and HERBICIDES, some medicines, and to preserve wood from rotting. When these products are made and used, arsenic is released into the environment. It builds up where certain metals are being made, where crops are sprayed with arsenic compounds, and where large amounts of fuel, especially COAL, are being burned.

People may inhale arsenic from polluted air, consume it in food or water, or be exposed to it in certain jobs. It can cause short- or long-term illness, involving damage to the skin, kidneys, liver, circulatory system, or nervous system. People who are in contact with arsenic products on the job have been shown to have higher than average rates of skin and lung cancer.

CADMIUM

Cadmium is used in coatings in metals; as a coloring agent in PLASTICS; in certain products made of polyvinyl chloride (for example, the white plastic "PVC pipe" used in household plumbing); and in nickel-cadmium batteries. Because it is present in such a variety of products, it is hard to control the spread of cadmium into the environment. It escapes during the production of metals and during the burning of coal and plastic. It occurs in rainwater, and in the RUNOFF from roads, mines, and industrial plants, and can build up in the sediments of rivers and bays.

Most of the ill effects of cadmium have been seen in people

Some Common Heavy Metals and Their Effects		
Metal	**Sources**	**Effects**
Arsenic (As)	Fossil fuels; smelting; batteries; pesticides; herbicides; mine tailings	Damages skin, liver, kidneys, circulatory or nervous system; carcinogenic
Cadmium (Cd)	Fossil fuels; zinc mining and processing; polyvinyl chloride; batteries; fertilizers; incineration	Damages lungs, kidneys, bones; carcinogenic
Lead (Pb)	Leaded gasoline; batteries; paint; smelting	Brain damage; hearing loss; behavioral disorders
Mercury (Hg)	Burning of coal; paint; fungicides; various industrial uses	Damages kidneys, nervous system; birth defects
Selenium (Se)	Electronics; glass manufacturing; photocopying	Damages lungs, kidneys, liver, spleen

whose jobs involve contact with it. It can cause severe damage to the lungs, kidneys, and bones. It is also known to cause cancer in animals. So far, the most dramatic case of poisoning from cadmium in the environment occurred in the 1950s and 1960s in Japan, where cadmium from a mine polluted the Jintsu River. Some people who used the river developed a painful and crippling disease of the joints and bones.

LEAD AND MERCURY

Lead is a very common and widespread heavy metal. It has been used by people for at least 4,000 years, and the danger of lead poisoning has been known for centuries. Today, it is mainly used in lead batteries but is present in many other products. Lead probably causes ill effects in more people than any other heavy metal.

Mercury is used in some batteries, in paint and seed grain to protect against mildew, and in other products. Mercury enters the environment as vapor, in paint dust, in mercury-treated grain, and in runoff from dumps and industries. Mercury has many toxic effects, including severe or fatal damage to the nervous system. Most poisoning occurs in people who work with mercury, or who eat mercury-treated grain by accident. However, people sometimes are poisoned by eating FISH or other animals that feed in polluted areas, and contain dangerous amounts of mercury in their bodies. The worst case of this type of mercury poisoning occurred in Minamata, Japan.

SELENIUM

Like other heavy metals, selenium is used in industry. It is used in the manufacture of electronics and glass and in photocopying. People who work with selenium may suffer a variety of health problems, including damage to the lungs, liver, kidneys, and spleen. However, selenium has another side: a very tiny amount of it is actually needed by the body to maintain proper health.

Selenium causes problems in the environment without much help from human polluters. It is naturally present in some soils, and is taken up by some PLANTS. These plants can then poison LIVESTOCK or people. On RANGELAND where selenium is present, cows or sheep may die of a form of selenium poisoning known as "blind staggers." In some parts of the American West, this problem has been made worse because people have brought new plants into the environment which are good at taking up selenium. [*See also* COPPER; FUNGICIDE; HAZARDOUS WASTE; MINAMATA DISEASE; MINING; and TAILINGS.]

Herbicide

▶ A substance used to kill unwanted PLANTS. Many plants are considered by people to be a nuisance. Such plants are often referred to as weeds. To get rid of weeds people often pull them up out of the ground. Sometimes

weeds are killed using chemicals called *herbicides*.

Herbicides are often used by farmers and homeowners as well as by companies and the government to clear power-line rights-of-way, railroad rights-of-way, highways, and other public areas of unwanted plants. Herbicides, also called DEFOLIANTS, have also been used in wartime to kill plants that could hide enemy soldiers. About 60% of the PESTICIDES used in agriculture in the United States are herbicides.

TYPES OF HERBICIDES

Herbicides are classified as either inorganic (having no CARBON) or organic (having carbon). Examples of inorganic herbicides include arsenic, COPPER, sulfur, and common salts.

Organic herbicides are divided into three groups according to how they affect plants. One group is contact herbicides, which kill plants on contact with their leaves. Atrazine and simazine are examples of

Major Types of Herbicides	
Type	**Effects**
Contact	Interferes with photosynthesis
Systemic	Stimulates excessive production of growth hormones; promotes inability to obtain sufficient nutrients to sustain accelerated growth
Sterilant	Sterilizes soil by killing essential microorganisms

contact herbicides. The second group is systemic herbicides, which enter the plant through its roots or leaves and move through the plant's circulatory system to other plant parts. Examples of these are 2,4-D and 2,4,5-T. The third group of herbicides is called the soil sterilants, which are used to kill organisms that live in the SOIL. They kill nematodes, FUNGI, weeds, and BACTERIA. An example is Dymid.

ACTION OF HERBICIDES

Some systemic herbicides are plant hormones. They can stimulate wild, rapid growth in some parts of the plant and inhibit growth in other parts. Other systemic herbicides interfere with PHOTOSYNTHESIS. Systemic herbicides have a long-term effect on plants. This contrasts with the effects of contact herbicides, which work very rapidly to kill the plant in a few days.

HERBICIDES AFFECT NON-TARGET ORGANISMS

Many herbicides affect organisms other than the weeds they are being used to kill. A herbicide containing 2,4,5-T was used during the Vietnam War as a defoliant on some FOREST areas. This herbicide is a component of Agent Orange. The preparation contained a chemical called DIOXIN. Dioxin causes birth defects and CANCER in animals, which are not the intended targets of herbicides. Because of dangers associated with its use, the herbicide 2,4,5-T was banned for use in the Vietnam War by the U.S. Department of Defense in 1970. At the same time, the U.S. Surgeon General announced that the U.S. government would limit the use of that chemical around homes, farms, and other areas.

PRECAUTIONS REGARDING HERBICIDE USE

Herbicides are very effective in killing plants. However, because of their potential harm to other organisms, they must be used with extreme care. Many of the effects on other organisms that might come into contact with herbicides are not yet known. As dangerous effects become known, the use of certain chemical products may be banned or severely restricted.

Very specific directions for use of herbicides by qualified persons are printed on herbicide labels. These directions must be followed to limit subtantially the possible harmful effects of these chemicals on the ENVIRONMENT.

INTEGRATED PEST MANAGEMENT

INTEGRATED PEST MANAGEMENT (IPM) is an approach to using an array of techniques to control unwanted plants and animals. It combines the benefits of chemical technology, such as herbicide use, with ecological pest management, based on knowledge of organisms. This approach uses all suitable methods to bring about long-term management of pest populations and has minimal environmental impact. [*See also* AGROECOLOGY; BIOLOGICAL CONTROL; HEAVY METALS POISONING; INSECTICIDE; PESTICIDE; and PEST CONTROL.]

Herbivore

▶ An animal that eats PLANTS. The term describes one of a series of organisms in the FOOD CHAIN, which consists of plants, herbivores, and CARNIVORES and OMNIVORES. Herbivores eat plants, digest plant tissues, and turn the chemicals in the tissues into their own nutrients. Carnivores eat herbivores and turn the herbivore tissues and chemicals into their own nutrients.

ADAPTATIONS OF HERBIVORES

Animals that eat plants need special ADAPTATIONS for chewing the tough material in plant cells. This material is called *cellulose* and is present in the outer layer of all plant cells. There may be additional material in plants that make them difficult to chew, such as lignin, cutin, and suberin.

◆ Two common herbivores are cottontail rabbits and while-tailed deer.

Some mammalian herbivores have molars, teeth with ridges for chewing. Seed-eating birds have tough beaks for breaking seed coats. Snails have a rasping structure in their mouths called a *radula* that enables them to eat plant tissue. The chewing mouth parts of a grasshopper are able to break up cellulose. [*See also* ENERGY PYRAMID; FOOD WEB; GRAZING; and PRODUCERS.]

Humus

▶ The soft, shapeless, usually dark-brown matter formed by the DECOMPOSITION of organic matter, found in healthy SOIL. Humus is important in maintaining good soil texture, the soil's ability to hold moisture, and, most of all, the soil's fertility.

Humus fills the spaces between the MINERAL grains or particles that make up soil. It also helps those grains or particles to cling together and form small crumbs. Crumbly soil makes it easier for plant roots and earthworms to travel. Earthworms mix and enrich the soil and enable air to circulate throughout it as they burrow. Humus stores vital nutrients and acts like a sponge, holding water near the surface, within reach of plant roots. Thus, humus increases soil's fertility.

Enormous populations of microscopic animals, PLANTS, and FUNGI are supported by humus, including the very ones that make humus as a by-product of their life processes. These organisms, called *saprotrophs,* use the material of dead plants and animals, as well as animal excrement, as their food source. As they digest the material, complex organic molecules are broken down into simpler compounds that may then be used by other organisms, such as plants. This process, called *humification,* is a vital step in the cycling of nutrients.

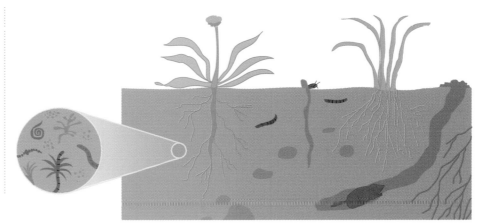

◆ Healthy soil contains saprotrophs like earthworms and fungi that help to decompose organic matter and turn it into humus.

◆ In spring, the soil is wet and rich in humus.

◆ Productive gardens such as these thrive in humus.

Humus is also important in soil and water CONSERVATION. Humus soaks up rainwater like a sponge and holds it for a time near the surface of the soil, where thirsty plant roots can absorb what they need. Without the humus, water would quickly evaporate, run off, or settle deeper than plant roots can reach. Without water, the plants could not live, and without the plants' strong networks of fine roots to hold TOPSOIL in place, soil EROSION would take place. The water that passes through the humus trickles down through deeper layers of soil and rock to form stores of groundwater.

By holding water, humus also helps prevent LEACHING, which is the removal of soil materials in a solution. Solutions are formed when particles of soil material, including mineral nutrients important to plants, are completely dissolved in water. When the water trickles downward, or percolates, to deeper levels of soil, the material dissolved in it trickles down as well, beyond the reach of the plants that need it.

PESTICIDES and other chemical contaminants may seriously damage or even destroy the communities of plants and animals that add humus to the topsoil. Other possible threats to the production of humus, and therefore to soil fertility, are CLEAR-CUTTING OF FORESTS, "clean" FORESTRY, and industrial and monocultural agriculture. These practices remove organic material and thus reduce the soil's humification potential. Plants will not grow in soil that lacks the rich plant food that humus provides. Soil has an average depth of only 2 to 3 feet (0.6 to 0.9 meters), and the fertile top layer is usually only inches deep.

Fertile topsoil is made by natural processes, which work very slowly. They may take as long as 600 years to produce only 1 inch (2.5 centimeters) of soil. But good topsoil can be lost through carelessness or abuse in a very short time. [*See also* BIOCHEMICAL CYCLE; CHEMICAL CYCLE; DECOMPOSER; DECOMPOSITION; DETRITUS; IRRIGATION; and WEATHERING.]

Hunter-Gatherer Society

▶ A group of people who meet their basic needs by obtaining food and other resources through fishing, HUNTING, and collecting materials from the ENVIRONMENT. The first humans may have appeared on Earth as many as 1.6 million years ago. To obtain the resources needed for their survival, these

◆ Although they make up only a small part of the total population of Earth, hunter-gatherer societies still exist in many remote parts of the world.

early humans spent much of their time wandering over the land looking for food and suitable natural structures, such as caves, that could provide shelter. Over time, these people began to build simple tools and weapons from rocks and other readily accessible materials that could be used for activities such as hunting and fishing. Groups of peoples who live in this way are called *hunter-gatherer societies*.

Before the AGRICULTURAL REVOLUTION that occurred about 10,000 years ago, virtually all humans existed in hunter-gatherer societies.

Such a lifestyle did not allow for rapid POPULATION GROWTH because many people died from starvation and diseases related to MALNUTRITION. In addition, populations were widely scattered, rather than concentrated in towns or cities.

After the agricultural revolution, the number of people living in hunter-gatherer societies decreased greatly. People began to gather together in small towns organized around farms that provided the foods the people needed to survive. Such societies allowed for an increase in population growth because the availability of food resulted in fewer deaths related to starvation and malnutrition.

Today, hunter-gatherer societies make up only a small part of the total human population on Earth. However, such societies do exist in underdeveloped nations and in remote parts of developing nations. They include the aborigines of Australia and the bushmen of southern Africa. [*See also* FRONTIER ETHIC and INDUSTRIAL REVOLUTION.]

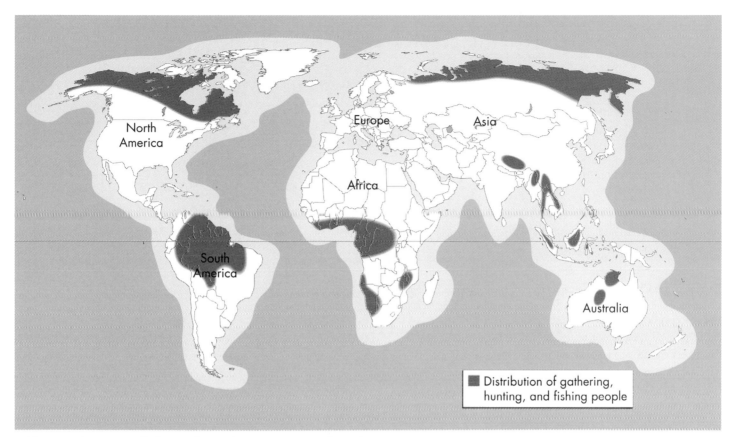

Distribution of gathering, hunting, and fishing people

◆ Hunter-gatherers usually live in the tundra or in equatorial rain forests.

Hunting

❱ The act of capturing or killing animal SPECIES. People hunt animals for a variety of reasons. Many animals are hunted for the food or the products they can supply, such as furs and feathers. ZOOS and other research organizations may hunt and capture animals so that the animals can be displayed for the public and studied to increase knowledge of the species. Such organizations may also capture animals to breed them and increase their population size. Many people also hunt animals for sport.

Most hunting is regulated by laws. For example, in the United States, it is legal to hunt certain species, such as deer, rabbits, ducks, and geese, at particular times of the year. This practice provides people with a sporting activity and helps prevent animal populations from becoming too large. To prevent overhunting, hunting organizations work closely with governmental agencies to make sure that only a certain percentage of these animals are killed each year.

Not all countries have the same hunting laws. Thus, animals that cannot be hunted in one country may be considered legal game in

◆ Deer are often hunted in many parts of the United States.

◆ People use many things for hunting, including rifles.

others. For example, many nations, including the United States, have agreed to stop the hunting of WHALES for commercial purposes. Other countries, such as Norway and Japan, have decided not to join in the ban because of the great economic benefits they receive from whaling.

In many countries that have wildlife laws, overhunting is no longer a major cause of EXTINCTION. However, in some areas of the world, illegal hunting, or POACHING, of certain species is a leading cause of extinction. Poaching is often done for sport. However, most poachers hunt animals for the profits they can get from the sale of the animal or its products. Skins, furs, feathers, and ivory tusks are only some of the animal products sold or traded illegally by poachers. Such practices have placed many types of animals, including GORILLAS, TIGERS, leopards, and ELEPHANTS, at the brink of extinction. [*See also* ENDANGERED SPECIES; NATIONAL WILDLIFE REFUGE; PET TRADE; and WILDLIFE.]

Hybridization

▶ The crossbreeding, or mixing of the GENES, of two different SPECIES that result in an **offspring** called a *hybrid*. A hybrid has traits of both species.

Hybridization changes the GENE POOL available to an organism. In some cases, these changes may strengthen the organism's chances of survival in a certain type of ENVIRONMENT. However, some genes from another species may be ADAPTATIONS to a particular environment or living conditions. Organisms having these characteristics may not survive in any other environment. Through NATURAL SELECTION, these organisms will be eliminated from unsuitable environments.

Hybridization can be a natural process. For example, the pollen from one type of willow tree can be transferred to the flower of another by an INSECT. The flower produces a willow seed that grows into a hybrid willow tree—one that carries genes from two different species of willow trees.

Animal hybrids are produced when two species mate with one another. Often, as in the case of mules, such breeding is controlled by humans. A mule results from the mating of a male donkey and a female horse. The mule is sterile; it cannot produce more mules. However, a mule is stronger and more resistant to disease than either of its parent species and so has a longer life expectancy. These traits make mules well suited to certain types of farm work.

Not all hybrids are sterile. The mating of black-headed grosbeaks and rose-breasted grosbeaks in the Kansas/Nebraska area has produced a **fertile** hybrid grosbeak. The feather color of the offspring of this hybrid grosbeak varies from that of the black-headed to that of the rose-breasted variety.

◆ A mule results from the mating of a male donkey and a female horse.

Ranchers often use hybridization to improve their LIVESTOCK. They have, for example, mated differing varieties and species of sheep and cattle to produce offspring that yield more meat and are resistant to disease. Artificial hybridization of PLANTS has produced many varieties of flowers and crop plants such as corn. Such hybridization may be carried out through the transfer of the pollen of one plant to another. In plants that reproduce by root rather than by seed, the stem of one plant is often attached or grafted to the root of another. [*See also* BIODIVERSITY; FLOWERING PLANT; GENETIC DIVERSITY; GENETIC ENGINEERING; GENETICS; and SPECIES DIVERSITY.]

Hydrocarbon

▶ A chemical substance that contains hydrogen and CARBON. Hydrocarbons occur naturally in various forms. For example, hydrocarbons make up FOSSIL FUELS, such as PETROLEUM and NATURAL GAS used to power motor vehicles, electrical plants, and factories. Hydrocarbons are also the main chemicals in natural rubber, an elastic substance obtained from certain tropical PLANTS. Hydrocarbons are also used in the making of synthetic substances such as lubricants, PLASTICS, synthetic rubber and fibers, explosives, and industrial chemicals.

Although hydrocarbons are naturally occurring substances, they are known to cause several envi-

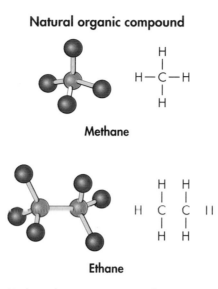

Natural organic compound

Methane

Ethane

◆ Hydrocarbons are compounds containing only hydrogen and carbon. Methane (CH_4) is the simplest type of hydrocarbon.

ronmental problems, particularly AIR POLLUTION. Hydrocarbons are released into the air whenever fossil fuels are burned. The thick black smoke seen emitted as exhaust from cars, trucks, and buses is hydrocarbon fuel that is not completely burned by the engine. In the air, hydrocarbons combine with sunlight and other chemicals to produce SMOG, the yellow-brown haze seen over many large cities. Smog reduces visibility and causes several health problems, including eye and lung irritation.

The burning of hydrocarbons may also contribute to the GREENHOUSE EFFECT. When fossil fuels are burned, CARBON DIOXIDE (CO_2) is released. As the most significant GREENHOUSE GAS, carbon dioxide contributes to the potential problem of GLOBAL WARMING.

To prevent some of the environmental problems caused by hydrocarbons, many scientists rec-

ommend the use of ALTERNATIVE ENERGY SOURCES, such as HYDROELECTRIC POWER, WIND POWER, and SOLAR ENERGY, in place of fossil fuels. Some countries, such as France, now obtain much of their ELECTRICITY from NUCLEAR POWER. Some scientists have suggested using ethanol, a less-polluting FUEL that is made from corn.

Hydroelectric Power

▶ The use of DAMS and RESERVOIRS to store water that is needed to generate ELECTRICITY. To generate electricity, the water must flow from a higher place to a lower place, such as from a waterfall or dam. The gravitational energy of the falling water (water dropping because of gravity) is then captured and used to generate power.

For centuries, people have dammed streams and rivers to hold back water. Early settlers built dams and small waterways to run water wheels that provided mechanical power for mills used to grind flour or saw logs. Later, people used water power to turn turbines to make electricity.

Modern dams are very large. They are capable of harnessing water that can be used to make electricity. Not all dams help produce electricity, however. Dams are also built to control flooding, to improve river navigation, or to hold water for IRRIGATION or drinking water supplies.

Hydroelectric projects produce electric power as falling water rotates a large turbine in a magnetic field. The amount of power produced is determined by the energy of the water turning the turbine. Water falling a long distance through a high dam releases more energy than water falling a short distance. Large amounts of water can create more electricity than smaller amounts of water. The water held by a dam can be thought of as a renewable source of energy because the water is not used up. The operation of a hydroelectric power station depends on a predictable water supply, and water in streams and rivers can be affected by drought and CLIMATE CHANGE.

TYPES OF HYDROELECTRIC POWER

Hydroelectric power is commonly produced by three different kinds of electrical generating systems—large-scale high dams, small-scale dams, and pumped storage systems. Large-scale high dams create large reservoirs on large rivers. High dams exist on many major rivers in the United States. Major high dam projects exist in the Tennessee River and in the Columbia-Snake River systems. The lifespan of high dams is uncertain because the reservoirs may fill with silt in as few as 30 or as many as 300 years. As reservoirs fill in, they store less water and are capable of generating less power. During this time, hydroelectric energy can become nonrenewable.

Small-scale hydroelectric plants, sometimes called *run of the river*

◆ Whitehorse Hydroelectric Dam, Canada.

◆ Karapiro Hydro Power Plant, New Zealand.

◆ Hydroelectric dams are found throughout the world.

plants, generate electricity based on the river flow. Usually, these small dams create small reservoirs and may not dam up a significant amount of water. These small plants can be affected by WEATHER cycles that create droughts.

Pumped-storage systems are systems that move water between two reservoirs. These systems usually have fossil fuel power-generating systems. The storage reservoirs are used to provide extra electric power during peak times of the day, usually in the morning and early evening. At times of peak power demand, water from an upper reservoir flows through turbines to make electricity. This water is then stored in a lower reservoir. At night, when power demand is low, electricity from a fossil fuel plant is used to pump water from the lower reservoir back to the upper reservoir. The depth of a pumped-storage reservoir changes, creating environmental problems like bank EROSION. Pumped-storage systems are more expensive to operate than conventional dam and reservoir systems.

HYDROELECTRIC POWER AROUND THE WORLD

In the United States, hydroelectric power supplies 7–15% of all electricity. Hydroelectric power accounts for about 21% of the world's electricity supply, but not every region makes use of hydroelectric power.

Building hydroelectric dams is more complex and expensive than just damming a river. Most hydroelectric power is produced in developed countries, especially those with large rivers and high mountains. For example, Austria, Switzerland, and Canada generate about 70% of their electricity using hydroelectric power. While many developing countries get up to 50% of their electricity from hydropower, they use less than 10% of their potential hydropower generating capabilities. On the other hand, the United States has placed dams on almost all of the sites that could generate hydroelectric power economically.

Many countries are developing large-scale hydroelectric projects to meet additional demands for electric power. Some of the projects are huge even by modern standards. For example, the James Bay project proposed for northern Quebec, Canada, would reverse the flow of

◆ The Shasta Dam in California is the second tallest in the world.

19 major rivers and flood an area the size of the state of Washington. In China, damming the Yangtze River for the Three Gorges Project will create the world's largest hydroelectric dam and reservoir, creating a 360-mile long (590-kilometer long) lake. It involves flooding farmland and hundreds of factories, requires the moving of 13 cities, 140 towns, and 326 villages, and displaces over 1 million people. Such huge projects can create many problems.

EFFECTS OF HYDROELECTRIC POWER DEVELOPMENTS

Although hydroelectric power is considered to have fewer adverse effects on the ENVIRONMENT than the production of electricity from fossil fuels and NUCLEAR POWER, river damming and reservoir building have a variety of effects both good and bad. Hydropower developments have produced relatively inexpensive electric power, and continued operation of power dams requires only a few people. Hydropower generating stations rarely have to be shut down.

Most industrialized countries have used most of the best sites for hydropower development. Building dams in developing countries poses new challenges. Large-scale projects flood river channels, eliminating HABITATS for NATIVE SPECIES. Changing the pattern of water flow also eliminates the replenishing of nutrients on river FLOODPLAINS, so agricultural production may be reduced. Often, dams have MULTIPLE USES. In the western United States, dam building has often been paired with the development of irrigation projects that reroute water and bring agricultural pollutants back to the main river. Water can actually be moved from one river basin to another, sometimes transporting unwanted SPECIES and often changing the availability of water in a formerly well-supplied area.

The development of high dams in tropical regions has led to the spread of disease as additional areas are flooded and irrigated. For example, the Aswan High Dam on Egypt's Nile River is credited with increasing the incidence of schistosomiasis (or bilharzia), a parasitic disease carried by snails whose populations expand in stagnant water.

Dams can also affect fisheries by altering water temperatures, changing the upstream food supply, or acting as barriers to the movement of migratory fish like SALMON even when FISH LADDERS have been built. While hydroelectric projects can create reservoirs that develop new fisheries and new recreation sites for people, they can also threaten fisheries, destroy WETLANDS, change scenic rivers, and alter natural aquatic communities. Excessive SEDIMENTATION caused by increased agriculture can threaten the lifespan of dams. Building dams in dry regions can produce excess evaporation of water, causing valuable water to be lost to the ATMOSPHERE. In some places, WASTEWATER is polluting the newly developed lakes. [*See also* ALTERNATIVE ENERGY SOURCE; AQUIFER; ARTESIAN WELL; BIOCHEMICAL OXYGEN DEMAND (BOD); BONNEVILLE POWER ADMINISTRATION; TENNESSEE VALLEY AUTHORITY; and THERMAL WATER POLLUTION.]

Hydrology

The study of the properties, distribution, and effects of water on Earth's surface, in the SOIL and underlying rocks, and in the ATMOSPHERE. Living organisms contain more water than any other substance. Every day, we drink water, wash in it, dispose of wastes in it, and rarely give it a second thought. When we take a closer look at it, however, we realize that this common substance has several uncommon properties. These properties are so important to organisms that we cannot imagine life existing on any planet that does not have an abundant supply of water. The main reason there is life on Earth while there is none on nearby planets is that water exists on Earth in the HYDROSPHERE. Hydrologists study the hydrosphere.

PROPERTIES OF WATER

The unique properties of water result from the structure of its molecules. A water molecule contains an atom of OXYGEN bound to two atoms of hydrogen. The molecule is electrically charged, causing water molecules to bond to one another. This structure gives water a number of properties that are vital to life.

1. Water sticks to itself and to other substances. You can fill a glass of water slightly above its brim. A mosquito larva can hang from the surface of a pond. Both are possible because of water's *surface tension*, which makes the surface of

◆ A large volume of Earth's water is in the polar ice caps.

water act as though it were covered by a skin. The molecules of water also stick to any electrically charged surface. This accounts for the *capillarity* of water—its ability to move upward, against the force of gravity, in small places such as the pores in paper or soil.

2. Water is a solvent. More things dissolve in water than in any other liquid. When a substance dissolves, its molecules separate from one another and mingle with molecules of the solvent. Because water is electrically charged, it dissolves other charged molecules such as salts, but it does not dissolve uncharged molecules such as oil. This means that water dissolves salts in the soil and sulfur compounds in the air to form ACID RAIN, but an oil spill floats on water and does not mix with it.

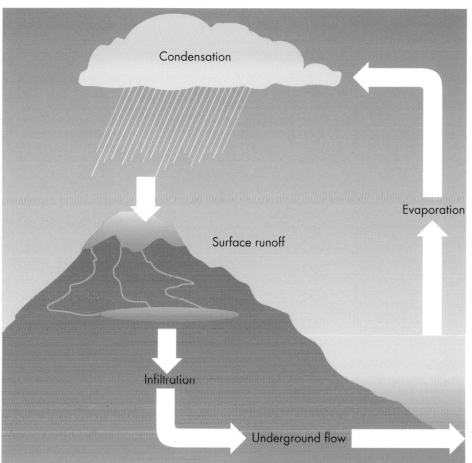

Condensation

Surface runoff

Evaporation

Infiltration

Underground flow

◆ In the water cycle, the sun's heat causes the water to evaporate and turn to water vapor. As it rises, the water vapor cools, condenses, and cycles back to Earth as rain or snow.

3. Water is a good evaporative coolant. It takes a lot of energy to make water molecules move fast enough to break the bonds that hold them together. When this does occur, liquid water becomes a gas, water vapor, in which each molecule is separate. Water vapor molecules carry a lot of heat energy away with them. Thus, when water evaporates from a body, it cools the body. This is why we sweat when we are hot and why we use large quantities of water as coolants in power stations and industrial processes.

4. Water is the only common substance that expands rather than contracts when it freezes. The low density of ice is the reason icebergs float. In temperate climates, this means that ice floats on top of water in winter, forming a blanket

◆ An artesian well produces free-flowing water under its own pressure in the ground.

◆ Water locked up as ice and snow melts when spring comes and flows into streams and rivers.

of insulation between the water and the cold air above. This slows the formation of more ice and protects organisms that live below the ice from freezing. Because water expands when it freezes, it also breaks pipes, cracks streets and rocks, and forces us to put anti-freeze in our cars.

DISTRIBUTION AND CONSUMPTION OF WATER

There is water, water, everywhere, but most of it is of no use to human populations. Most of the water on Earth is salt water in the OCEANS. Of the 3% that is fresh, most is locked up in polar ice caps and GLACIERS. Most of the rest occurs under-ground. Less than one-hundredth of 1% of the water on Earth exists in the atmosphere, rivers, and lakes where we can get at it.

Most of the water used by households and industry is eventually returned to the water supply. It is said to be withdrawn, but not consumed. In all countries, how-ever, much more water is used to irrigate farmland than goes to households, industry, and all other uses combined. Furthermore, the water withdrawn for IRRIGATION is less likely to return to the water supply.

Less than one-sixth of the water withdrawn for households and industry is consumed. More than half of the water withdrawn for irri-gation never returns to the water supply because it evaporates into the air.

We continually pour pollutants into water, so how is it that all the fresh water on Earth is not already too polluted to drink? The answer is that water vapor, the gas that evap-orates from a damp surface into the air, is pure water. The pollutants are left behind as the water evaporates.

Water is continuously purified by the WATER CYCLE, in which water moves between the atmosphere, the land, and the oceans. The water cycle is driven by the sun's heat energy, which causes water to evaporate. Some of the water evap-orates from soil, roads, and lakes. A lot is evaporated from PLANTS, most of it during PHOTOSYNTHESIS.

Seventy percent of Earth's sur-face is ocean, and this is where most water evaporates. Much of the water that evaporates eventually descends on the ocean as rain or snow. Only about 30% falls on land, where it drains slowly back toward the ocean as RUNOFF. This runoff makes up our "water income," the recycling supply of purified water upon which life depends. [*See also* ACID RAIN; AQUIFER; ARTESIAN WELL; HYDROELEC-TRIC POWER; PRECIPITATION; RESERVOIR; SURFACE WATER; and WATER, DRINKING.]

Hydrosphere

❙❭ All the water on Earth. The water on Earth exists on and in the

♦ Automobiles produce carbon monoxide that can cause hypoxia.

ground and the ATMOSPHERE. The *hydrosphere* ("water sphere") refers to the water near Earth's surface, not the water in the atmosphere.

The hydrosphere includes gaseous, liquid, and solid forms of water vapor, liquid water, and ice. It includes the OCEANS, seas, lakes, rivers, and streams, as well as water in underground AQUIFERS. The hydrosphere also includes the layer of ice on a pond in winter, as well as huge bodies of ice such as icebergs floating in the ocean. In addition, polar ice caps cover large areas at the North and South Poles. These masses contain more fresh water than all the lakes and rivers in the world. Some of the ice in the hydrosphere lies underground in the form of *permafrost*, the permanently frozen soil water found in TUNDRA ECOSYSTEMS.

Hypoxia

▌A deficiency in the amount of OXYGEN reaching the body's tissues. Hypoxia may result from physiological problems or environmental factors.

One physiological problem that may lead to hypoxia in humans is anemia. Anemia is a condition in which the body does not produce enough red blood cells. Red blood cells contain a protein called *hemoglobin*. The job of hemoglobin in red blood cells is to carry oxygen from the lungs to the cells and tissues of the body. If fewer red blood cells are present in the body, the amount of oxygen that can be carried to body tissues decreases.

There are two common environmental factors that lead to hypoxia. The first is a decrease in the amount of oxygen present in air at high altitudes, such as mountaintops. With less oxygen in the air, the amount entering the lungs when a person inhales is reduced. As a result, less oxygen is carried to the body's tissues. The second environmental factor that leads to hypoxia is the presence of pollutants or poisons in the air. One of the more common pollutants or poisons that causes hypoxia is CARBON MONOXIDE.

Carbon monoxide is a gas that is given off when FOSSIL FUELS, such as oil and gasoline, and cigarettes are burned. Like oxygen, carbon monoxide can be carried by the hemoglobin of red blood cells. If a red blood cell carries carbon monoxide, it cannot also carry oxygen. Thus, cells and tissues that receive carbon monoxide from red blood cells are deprived of oxygen.

Hypoxia is also a term to describe the condition of a pond, lake, or region of the OCEAN containing insufficient DISSOLVED OXYGEN to sustain life. [*See also* AIR POLLUTION; BIOGEOCHEMICAL CYCLES; RESPIRATION; and SICK BUILDING SYNDROME.]

I-J

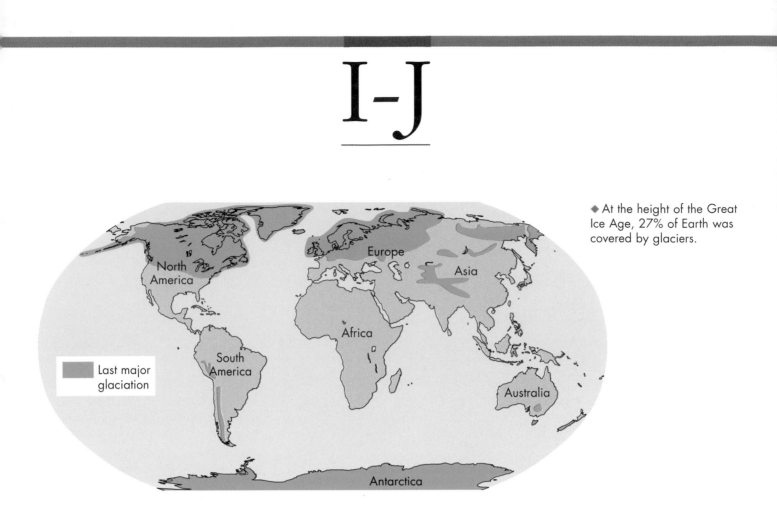

◆ At the height of the Great Ice Age, 27% of Earth was covered by glaciers.

Last major glaciation

Ice Age

▐▶Geological period characterized by the presence of ice sheets on vast portions of the continents. Ice ages, or glacial periods, have occurred several times during the past two billion years of Earth's history. The earliest known ice age occurred nearly 600 million years ago during the Precambrian era.

The most recent ice age began in the Pleistocene epoch two million years ago and lasted until 11,000 years ago. Known as the Great Ice Age, about 27% of Earth's land surface was covered by thick sheets of ice. In North America, for instance, ice sheets completely covered Canada and moved as far south as

New Jersey. In the Midwest, glaciers reached down into what is now St. Louis, Missouri.

Other parts of the world were covered by ice, too. In Europe, ice covered Scandinavia and reached as far south as Germany and Poland. Glaciers also covered the northern parts of Russia and Asia.

The gradual movement of GLA-CIERS can drastically reshape the surface of the land. The parts of the world once covered by glaciers still show scars of the last ice age. For instance, the GREAT LAKES were carved out of rock by glaciers that covered the area from about 250,000 to 11,000 years ago. Some parts of the world are still covered by glaciers today.

Scientists are not sure about which events trigger the severe changes in CLIMATE that cause ice ages. However, some scientists suspect that

changes in Earth's orbit around the sun may be related. [*See also* CLIMATE CHANGE; EROSION; and WEATHERING.]

Indigenous People

▐▶People living in an area where their cultural group originated or where they have existed for many generations. There are between 4,000 and 5,000 different indigenous, or native, groups in the world. Indigenous people live on every continent except ANTARCTICA.

Cultural influences from non-indigenous peoples have changed the basic way of life for most of the world's indigenous groups.

◆ Many indigenous peoples, such as the Inuit of Alaska and the Masai of East Africa, live in remote areas of the world where technology, trade, and other cultural influences have been slow to penetrate.

However, many indigenous cultures still exist in more remote areas like tropical RAIN FORESTS and DESERTS where they continue to practice their traditional ways.

The United States has hundreds of indigenous cultures, such as the Cheyenne of the upper Midwest, the Cherokees of the Southeast, the Hopis of the Southwest, and the Inuits of Alaska. Many Native American groups have lived in North America for thousands of years and possess their own languages and unique cultural practices. The lifestyles of many Native American groups have been altered by the spread of nonindigenous people; however, many of these groups still practice their traditional economies based on HUNTING and gathering, fishing, and small-scale agriculture.

THE FUTURE OF THESE GROUPS

Today, there is great concern about the future of indigenous cultures. Due to the expansion of technology and the growth of the human population, many indigenous cultures are being displaced from their homelands. In the South American rain forests, for example, traditional forest-dwelling peoples are forced to compete for space with farmers and developers who want the land for agriculture, roads, DAMS, and buildings.

Scientists fear that the loss of indigenous peoples in South America and other parts of the world may also mean a loss of knowledge about BIODIVERSITY. Many of the world's indigenous groups, such as those in the tropical rain forests of South America, live in areas with rich SPECIES DIVERSITY. These groups rely on a great many animal and PLANT SPECIES for survival. Researchers often consult with members of these groups to learn about the location and uses of plants and animals.

Industrial Revolution

▶ A period of great and sudden change in the economic system of Europe in the late eighteenth century and of the United States in the nineteenth century. During the Industrial Revolution, mechanized production quickly replaced an economic system based primarily on agriculture and home manufacturing.

The Industrial Revolution began in Europe in the 1700s. The period was marked by a number of important social changes that occurred over decades. The changes that took place produced a society that began to resemble modern, urban settlements.

Before the Industrial Revolution, most people lived on small farms or worked for large landowners. Families produced most of the things they needed from food to clothing to furniture. Some manufacturing took place in the cities. Most of the wealth and political power was held by a few.

◆ Native North Americans create hundreds of types of baskets from natural materials.

◆ The smoke from coal-burning factories caused severe air pollution problems in cities in the 1900s.

Technological improvements and new sources of power at this time brought about great changes. Agricultural developments made farming more profitable for large landowners. The large landowners bought out the small farmers, leading to an increased movement to the cities.

New inventions to increase textile production led to bigger factories that employed more people. New power sources were needed for the new machinery—COAL and steam provided cheap and efficient power. Factories replaced domestic labor systems of production. Machines such as the steam engine revolutionized cotton manufacture, and cotton production tripled.

As all of these developments were taking place and more and more people were moving to the cities, other changes were also occurring in the way people lived. Cities became crowded with a shortage of housing. Inadequate water and sewer services led to the rapid spread of diseases and WATER POLLUTION. The smoke from the coal that was being burned in greater and greater amounts led to AIR POLLUTION.

Some people feel that the changes that came with the Industrial Revolution led to great hardships in the way people lived and in its impact on the ENVIRONMENT. Others feel that the changes, although not perfect, led to a better way of life for most people.

By the mid-1800s, debates were beginning about how wealth should be distributed. Rival economic theories began to appear. Darwin's *On the Origin of Species* suggested that COMPETITION and the struggle for existence were fundamental. This belief raised the idea of free competition to high ground. Competition became so fierce that workers' wages were severely limited to the point that factory legislation was passed and trade unions were created to protect the workers.

At the time of all this thinking about production and the distribution of wealth, human populations were growing rapidly. Simultaneously, the number of farmers and the number of acres farmed were dropping. The more recent AGRICULTURAL REVOLUTION saw communities move away from common-field methods of production. Fences were built and small farms were consolidated into larger ones. At the same time, iron production doubled, and great canals and factories were built. Economic changes continued to stimulate commerce and alter the way wealth was distributed. Farmers gained wealth, and the earlier class structure began to deteriorate. At the same time, the wages of the laborers were dropping, and the access of these people to formerly common property disappeared.

These changes in the way people worked and the way wealth

was distributed remain with us to this day. Historian Arnold Toynbee has written that the Industrial Revolution shows that "free competition may produce wealth without producing well-being." Although the economy of most western countries is still based on the model developed during the Industrial Revolution, significant social changes have occurred, regulating business, industry, and commerce. In addition, the Industrial Revolution led to the possibility of the highly developed technology available today to a wide variety of people of many economic levels. [*See also* ACID RAIN; FOSSIL FUELS; FRONTIER ETHIC; GREENHOUSE EFFECT; HEALTH AND DISEASE; and TRAGEDY OF THE COMMONS.]

Industrial Waste Treatment

❙Techniques for disposing of any waste materials created by industrial activities. This includes the approximately 15% of industrial wastes that are hazardous and pose a threat to human health and/or the ENVIRONMENT.

Almost every stage of an industrial process produces some type of waste. This waste ranges from harmless materials such as sand to toxic chemicals such as DIOXIN. The careful and appropriate disposal of harmful wastes is a major problem for industry and a major concern for environmentalists.

Potentially dangerous wastes are common in industries such as oil refining, MINING, **smelting**, and power plants. Such wastes are also produced by companies that manufacture products such as PESTICIDES, PLASTICS, paper, paint, chemical **solvents**, batteries, medicines, clothing and textiles, leather, and explosives. Harmful wastes may exist as liquids, solids, or gases. Many are contained in SLUDGE that is poisonous, flammable, corrosive, infectious, or radioactive. Among the many materials classified as HAZARDOUS WASTES are those that may ignite or explode when exposed to water, air, or static ELECTRICITY. Other hazardous wastes are toxic to humans and/or other organisms.

If disposed of improperly, hazardous industrial wastes can pollute SOIL and water supplies. Animals and humans that eat food grown in the contaminated area may develop CANCER or other diseases, or may die from consuming the wastes. Unfortunately, some industrial wastes, such as LEAD, MERCURY, and other heavy metals, remain dangerous whenever levels stay high.

HAZARDOUS WASTE DISPOSAL

Improper disposal of hazardous industrial wastes has produced thousands of sites in the United States that threaten the environment and human health. The U.S. Office of Technology Assessment (OTA) estimates that 80% of the 240 million tons (218 metric tons) of hazardous waste created each year winds up in LANDFILLS. The LOVE CANAL disaster of New York State illustrates the danger of burying

hazardous industrial waste in a landfill.

In the late 1940s and early 1950s, more than 21,000 tons (19,110 metric tons) of industrial chemical wastes were dumped in the abandoned Love Canal landfill. A housing development and elementary school were built on the old landfill site in the mid-1950s. Soon after, chemicals started to seep out of the landfill and into nearby soil and water. Over the next 20 years, residents of Love Canal suffered from a variety of health problems including increased occurrences of cancer, miscarriages, birth defects, and disorders of the nervous system. The basements of homes flooded with corrosive sludge. Yards developed holes that filled with foul-smelling, highly acidic liquid.

An official investigation in 1977 uncovered a vast leak contaminated with toxic chemicals coming from the canal. In August 1978, New York State condemned the property and evacuated 240 families from Love Canal. Soon, Love Canal became a ghost town.

In April 1980, Congress set up the COMPREHENSIVE ENVIRONMENTAL RESPONSE, COMPENSATION, AND LIABILITY ACT (CERCLA). The act is also known as the SUPERFUND. The law authorized the ENVIRONMENTAL PROTECTION AGENCY (EPA) to clean up an estimated 22,000 abandoned HAZARDOUS WASTE sites in the United States. Included among them were 24 sites thought to be even greater threats to public health than the Love Canal. The cleanup is still going on today.

Since 1980, most of the advanced nations have adopted laws

regulating hazardous waste disposal. However, much of the hazardous industrial waste created in developed nations is still disposed of improperly. Often, these wastes are shipped to nations that have no regulations controlling disposal of hazardous wastes.

PROPER DISPOSAL

The EPA regulates the design and operation of landfills specifically designed for disposal of hazardous wastes, under the RESOURCE CONSERVATION AND RECOVERY ACT (RCRA) of 1976. Such landfills must isolate hazardous wastes from the environment, control potential LEACHING problems, and provide long-term monitoring. It is hoped these methods will assure prevention of future problems.

To isolate hazardous materials, the bottom of a landfill must have plastic or clay liners and layers of sand to keep the materials from reaching the soil. To handle these hazardous liquids, a landfill must have a leach collection system—

◆ Old oil drums were simply left on a waste site in the San Francisco Bay area without any protective devices used in modern landfills.

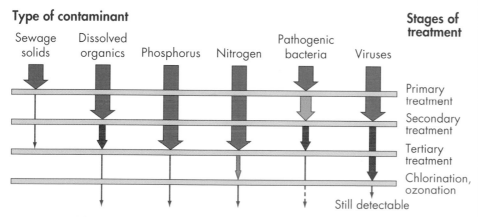

◆ Water used for industrial purposes is often degraded during use by the addition of solids and other wastes. By law, these waters must be treated before being released back into the environment. Each stage of treatment removes specific contaminants. The relationship between each stage and level of removal is shown here.

THE LANGUAGE OF THE ENVIRONMENT

polymer a compound, like plastic, consisting of many molecules.

smelting the heating of a material or ore to a high temperature to melt it or fuse it.

solvents substances capable of dissolving other substances.

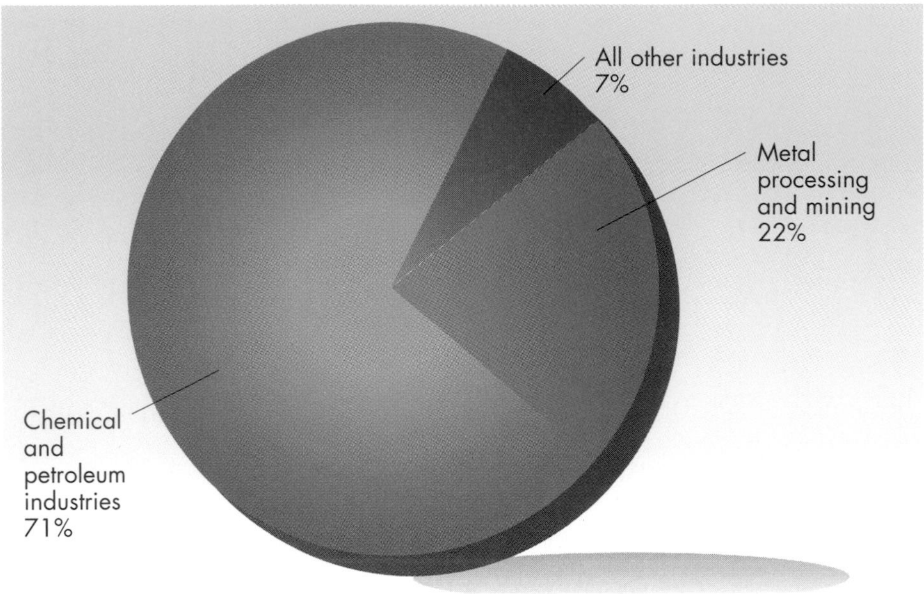

◆ Chemical and petroleum industries produce 71% of U.S. industrial wastes. The metal processing and mining industries account for 22%.

perforated pipes into which leached fluid can accumulate for removal. The completed landfill must be topped with a final vinyl liner, a layer of gravel, and a layer of TOPSOIL and vegetation.

The EPA believes no liner is capable of keeping all hazardous liquids out of the ground forever. So liquids are usually solidified to prevent leaking problems. A cement-based process turns inorganic liquids into rocklike pillars for burial. Organic materials may be mixed with a **polymer** to create a spongy mass that traps solid particles inside. Some materials may be bonded and coated with plastic to form solid bricks.

The incineration of organic hazardous waste greatly reduces problems with groundwater contamination. However, the burning process can pollute the air, unless control devices are used to capture PARTICULATES and remove toxic chemicals.

RECYCLING INDUSTRIAL WASTES

Heat produced by incinerating hazardous wastes gives off steam that can be used to generate electricity. Lead from car batteries can be saved for use in the manufacturing of new batteries. The recycling of other wastes such as iron, steel, ALUMINUM, and plastic can save energy, manufacturing costs, and needed space in landfills. [*See also* COGENERATION; HAZARDOUS SUBSTANCES ACT; HAZARDOUS WASTE MANGEMENT; HAZARDOUS WASTES, STORAGE AND TRANSPORTATION OF; HEALTH AND DISEASE; HEAVY METALS POISONING; MINAMATA DISEASE; PCBS; RADIATION; RADIATION EXPOSURE; RADIOACTIVE WASTE; RADIOACTIVITY; RECYCLING, REDUCING, REUSING; SOLID WASTE; TOXIC WASTE, INTERNATIONAL TRADE IN; WASTE MANAGEMENT; and WASTE REDUCTION.]

Infant Mortality

The death of children younger than one year of age. The death rate among infants younger than one year is higher than that for any other age group under forty.

There are many causes of infant mortality. Infant mortality often results from developmental defects that result from the GENES a child inherits from one or both of its parents or from trauma or other problems that occur during the development of an embryo or fetus. Infant mortality may also result from

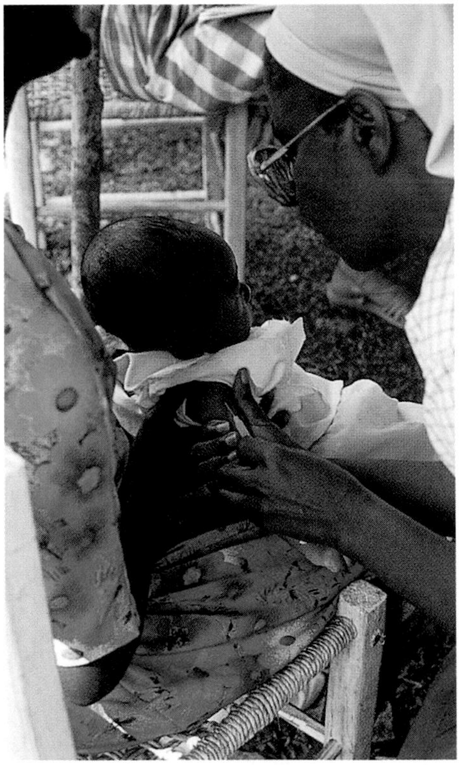

◆ Inoculating infants against childhood diseases helps reduce the incidence of infant mortality both in developing and industrialized countries.

trauma that occurs during the birth process or from diseases a newborn child or infant develops early in life.

In general, infant mortality rates are higher in developing countries than in industrialized nations. In developing countries, infant mortality often results from starvation or disorders related to MALNUTRITION. Infants in developing nations are also prone to diseases caused by poor sanitary conditions such as inadequate SEWAGE treatment and disposal methods and drinking water contaminated by PATHOGENS. A lack of access to medical care also contributes to the high infant mortality rates in some developing nations.

A significant number of infant deaths among humans can be prevented. For example, infant mortality is higher than average when an expectant mother does not get proper medical care during pregnancy, does not eat healthful foods, or uses harmful substances such as drugs, alcohol, and tobacco. Once the baby is born, the chances of survival can be increased through proper nutrition and regular medical care, including vaccinations designed to prevent the development of certain diseases.

Infant mortality is sometimes used as a measure of how well a society takes care of its people. Among the industrialized nations of the world, the United States has the highest infant mortality rate. Some people believe this situation results from the fact that not all citizens of the United States are provided with health care. [*See also* HEALTH AND DISEASE and HEALTH AND NUTRITION.]

◆ Irrigation water must infiltrate the soil in order to become accessible to roots.

Infiltration

The downward movement of water in SOIL. Water from rainfall or melting snow moves downward through soil and underground rock layers into what is called the ZONE OF SATURATION, where it forms groundwater.

Water from infiltration can also move through the soil into small streams. Because water is able to

◆ Soil contaminants can be carried to the groundwater by infiltration.

dissolve and transport substances, many contaminants in the soil or in rainwater may also be carried into SURFACE WATER or groundwater. If contaminants like PESTICIDES or heavy metals are in the soil, they can be carried to the groundwater by infiltration, and groundwater pollution can result. Various land uses, particularly agricultural and forestry practices, can significantly—and often adversely—affect infiltration. [*See also* AGRICULTURAL POLLUTION; HYDROLOGY; LEACHING; WATER CYCLE; and WATER POLLUTION.]

Insect

▶ A small, invertebrate animal with an external skeleton, three pairs of legs, and usually two pairs of wings. Most insects undergo metamorphoses, or changes in form with growth, within their life cycles. There are three times as many kinds of insects as there are all other kinds of animals combined.

Approximately 750,000 SPECIES of insects have been discovered. Although no one knows exactly how many insect species exist, some scientists estimate that many thousands or millions of undiscovered species remain to be found. The EVOLUTION of insects has taken many different paths. Their small size and the ability of most species to fly allow insects to occupy a wider variety of HABITATS than most other animal groups.

INSECTS AND THE FOOD CHAIN

Insects occupy every consumer TROPHIC LEVEL of the FOOD CHAIN. Thus, they play important roles in maintaining healthy ECOSYSTEMS. Many insects are primary CONSUMERS, eating PLANTS. Others are secondary consumers and eat primary consumers. Examples of insects that are primary consumers are aphids and **scale insects**. Both feed directly on plant tissue. Almost every kind of plant is eaten by some kind of insect. Insects may specialize, eating only one plant organ, such as the leaf or root. For example, aphids have long piercing and sucking mouthparts which they insert into the circulatory system of a plant to drink the plant juices, much as you would drink soda through a straw. Some of the scale insects, such as the armored scales, injure plants by sucking sap. They may be

◆ The horn of the rhinoceros beetle is a projection of its external skeleton. In beetles, the outer pair of wings becomes thickened and is used for protection.

so numerous that they kill the plant. Other herbivorous insects, such as caterpillars (the larval stage of butterflies), leaf beetles, and grasshoppers chew leaves. Some beetle larvae (grubs) feed on roots. Certain weevil and moth larvae feed on fruits.

Thousands of different insects are CARNIVORES, which are animals that eat other animals. Carnivorous insects may be secondary or tertiary consumers. Insects such as ladybugs, which eat aphids, are secondary consumers. They feed directly on the primary consumers, which eat plants. Digger wasps are also secondary consumers that feed on other insects or spiders. Adult digger wasps usually lay eggs on the prey's body. When the eggs hatch, the hatchlings feed on the prey insect or spider. Many secondary consumer insects, such as lice and fleas, feed on VERTEBRATES.

An example of a tertiary consumer is a trigonalid wasp whose larvae, or young stage, eat the larvae of other wasps which feed on caterpillars. The trigonalid wasp is a parasite that feeds on another parasite. Such organisms are called *hyperparasites.*

PREY, PREDATORS, AND PARASITES

A prey animal is one that is eaten by another animal. A PREDATOR is an animal that eats another animal. A parasite is an animal that lives in or on another animal called the *host.* Usually, the parasite harms, but does not kill, its host.

Many insects are prey to BIRDS, such as the nighthawk or purple martin. Nestlings often eat their weight in insects daily. Many fresh water FISH eat insects such as mayflies, stoneflies, caddisflies, mosquito and midge larvae, and the larvae of some aquatic beetles. Other insect-eating vertebrates include frogs, lizards, bats, skunks, moles, and shrews.

Some insects are predators that capture other animals. Usually, predators are secondary consumers or tertiary consumers. Ladybugs are predators, feeding on aphids and scale insects. Ground beetles also feed on other insects. Some ground beetles even feed on snails. Tiger beetles feed on a variety of small insects. The larvae live in burrows in the SOIL and ambush their prey

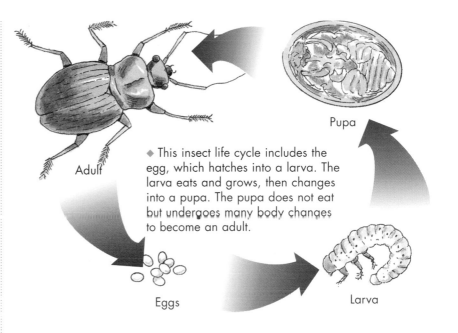

◆ This insect life cycle includes the egg, which hatches into a larva. The larva eats and grows, then changes into a pupa. The pupa does not eat but undergoes many body changes to become an adult.

as it passes by. Antlion larvae, called *doodlebugs,* also dig pits. They wait in the pit and feed on ants and other insects that fall in.

Chewing lice are external parasites of birds and MAMMALS. These insects cause skin irritations in their hosts. Fleas and bedbugs can also cause irritations in the skin of their hosts. Many larval stages of flies are internal parasites of humans and other animals, causing a condition called *myiasis.* The larvae of the ox warble fly live under the skin of

◆ A few kinds of butterflies migrate to warmer regions in the winter. The monarch may travel up to 2,000 miles (3,200 kilometers) from Canada to Mexico.

◆ The lubber grasshoppers are herbivores and have biting and chewing mouthparts.

◆ The larvae of any one of several hawk moths are called *hornworms*. Hornworms are herbivores. The tomato hornworm feeds on tomatoes growing in a field.

oxen. The sheep bot fly lives in the nasal passages of its host. Female parasitic wasps deposit their eggs directly into the larva of another insect (the host). When the eggs hatch, the young wasp larvae eat the host larvae before they can develop into adults.

SCAVENGERS

Many insects are scavengers. Scavengers feed on dead plants and animals and animal wastes. Wood-boring beetles, termites, and carpenter ants eat the tissue of fallen trees. In this process, the insects help break down the dead trees to form soil. Blowflies lay eggs in the bodies of dead animals. When the eggs hatch into larvae, they feed on the animal remains. Blowfly larvae can eat a horse faster than a lion can eat a horse. Other insect scavengers include burying beetles, which excavate the ground beneath the dead body of a small animal, let it sink into the hole, and then lay eggs in it. Adults and larvae then feed on the body.

INSECTS AND PEOPLE

Insects have existed on Earth for more than 300 million years. Humans, however, have existed only a little more than five million years. Insects exist in almost every type of habitat and occupy almost every kind of NICHE. They live in all land BIOMES, including GRASSLANDS, FORESTS, DESERTS, and even on the ice caps. Some live in freshwater lakes and ponds. Some live in WETLANDS. Some live in the soil. Some live inside plants.

Insects and the Human Food Supply

One of the ways insects affect human activities is by eating the plants we grow as food. In a natural biome, insects generally eat no more than 15% of the plant material. However, when humans take over an area for the growing of crops such as corn, wheat, rye, rice, and millet, grass-eating insects are provided with an almost limitless food supply. As a result, such species flourish. For example, the corn-borer may be provided with acres and acres of corn to eat. The cotton boll weevil has the best of all possible worlds in a cotton field. Some insects that tunnel into fruits are the codling moth, oriental fruit moth, apple maggot, plum curculio, and the nut weevils.

Some insects that feed on crops in the field by boring into their stalks and stems are the corn borer, wheat stem sawfly, and squash vine borer. Roots of crops are attacked by the larvae of such insects as the wireworms and white grubs, the corn rootworm, and the onion and cabbage maggots. Some feed upon seeds before they are planted. Some weevils and some moth caterpillars are seed-eaters. After the grain is processed into flour, the flour beetles begin eating. The

◆ The silkworm moth makes silk by spinning a cocoon, which is the silk.

extensive damage caused by many insects has led many farmers to use all means possible to prevent the wholesale destruction of their crops.

Biological Controls

Some modern farmers use BIOLOGICAL CONTROLS instead of or in combination with PESTICIDES to control unwanted insect populations. Biological controls have several advantages. They normally affect only the species intended. They are not toxic to other species, including people. Once a population of control species is established in an area, it usually reproduces itself.

Predatory insects can be used as biological controls. In one area, Australian ladybird beetles were brought into California orange and lemon groves to eat the fluted scale insect that was destroying the trees. Spiders eat insects. Leaving some weeds around soybean and cotton fields provides habitats for wolf spiders, which eat many kinds of insects. Another biological control method involves sterilizing male insects so they cannot fertilize the females. This technique is used for insect species in which the female mates only once in her life.

Bacteria can also be used for pest control. It infects the pests and reduces their number. Birds eat insects, so farmers can provide habitats and nesting sites that attract woodpeckers, purple martins, chickadees, barn swallows, and other insect-eating species. Some nematodes, worms that live in the soil, are used to kill pests. They are wrapped in a capsule with specific insect attractants and other chemicals to help them enter the pest's body and feed on its tissues.

Insects and Human Health

Insects can be an annoyance when they bite or sting. They may also transmit disease-causing PATHOGENS through their bite. Examples of insects that transmit disease include biting and sucking lice, fleas, flies, and mosquitos. Lice transmit typhus. Fleas transmit the BACTERIA that cause bubonic plague. Flies transmit the protist that causes African sleeping sickness. Mosquitos transmit malaria protozoa and typhoid fever bacteria.

Beneficial Insects

Although some insects do harm to humans directly and indirectly, we are dependent upon others for our food supply. It has been estimated that one-third of all of our food depends directly on insect pollinators. Many plants used as food are pollinated by insects. POLLINATION is the transfer of pollen from the male sex organ of a flower to the female sex organ of a flower. The pollen then grows and produces the sperm that is needed to fertilize eggs to produce the next generation of plants, through seeds. Unfortunately, the use of pesticides has killed not only insects that eat our food plants, but also insects that pollinate our food plants. This unwanted effect of pesticides has caused some farmers to have to buy bees and bring them to their farms to pollinate their crops.

Commercial Products Obtained from Insects

Honey and beeswax are used as food and in the manufacture of many products. Beeswax is used in making such products as candles, sealing wax, polishes, and certain kinds of inks.

Silk is obtained from a species of moth. An immature stage, called the *pupa*, spins a cocoon around itself made of a single thread of silk about 1,000 yards long. It takes about 3,000 cocoons to make a pound of silk. The silk industry is an ancient one that still exists in Japan, as well as in China, Spain, France, and Italy.

Shellac is produced from the secretions of the lac insect, a type of scale insect that occurs on figs and other plants in countries such as India and Sri Lanka. The lac scale forms crusts on twigs of the host plant, which are harvested and processed for the manufacture of shellac.

Cochineal dyes are made from scale insects that feed on cacti. The dye is purple and is made from the dried insects. Other types of dyes have been made from different scale insects and from galls produced by insects.

Insects, such as blister beetles, have been used in medicine. An extract is used in the treatment of some **urogenital** conditions. Bee venom has been used to treat arthritis. Blow fly larvae have been used to treat decaying tissues of some wounds in order to promote healing.

People eat many different kinds of insects. Locusts are eaten in some Arab countries. Some Africans eat ants, termites, beetle grubs, caterpillars, and grasshoppers. A type of caterpillar called *Gusanos de Mag-uey* is considered a delicacy in Mexico. [*See also* INSECTICIDE.]

Insecticide

▶A chemical substance that kills insects. Control of insects is necessary because they transmit diseases and damage crops.

The three major types of insecticides are stomach poisons, contact poisons, and fumigants. Stomach poisons kill when they are eaten. Contact poisons kill when they penetrate body coverings. Fumigants act when they are inhaled. Most insecticides are sprayed or dusted onto plants or into contained areas. Before the 1940s, most insecticides were made from toxic compounds of metals like COPPER or LEAD.

In the mid-1940s, highly effective insecticides such as dichlorodiphenyl trichloroethane (DDT) became available. DDT is a contact poison that can kill a wide variety of pests by disorganizing their nervous systems. During and after World War II, DDT helped stop the spread of typhus, plague, malaria, and yellow fever by killing the insects that transmitted the diseases. It also saved many valuable crops.

Although insecticides like DDT are useful, they have two major disadvantages. First, they are an environmental hazard. They can remain and build up in the ENVIRONMENT and harm people and WILDLIFE. Second, insects often develop resistance to insecticides. Insects that are resistant to a particular insecticide can tolerate the poison without becoming sick. As a result, the insecticide becomes less and less effective over time. Because of the adverse effects of DDT on wildlife, the United States banned its use in 1972. However, it is still used in some parts of the world, especially where mosquito-borne diseases are a serious problem.

Farmers and pest insect controllers now use fewer insecticides. They are turning to less harmful methods, such as BIOLOGICAL CONTROL and INTEGRATED PEST MANAGEMENT (IPM). [*See also* BIOACCUMULATION; CARSON, RACHEL LOUISE; and PESTICIDE.]

Major Types of Insecticides	
Type	**Examples**
Stomach poisons	Arsenic compounds
Contact poisons	DDT and other chlorinated hydrocarbons; pyrethrum and other extracts from flowers
Fumigants	Hydrogen cynanide, carbon disulfide, carbon tetrachloride

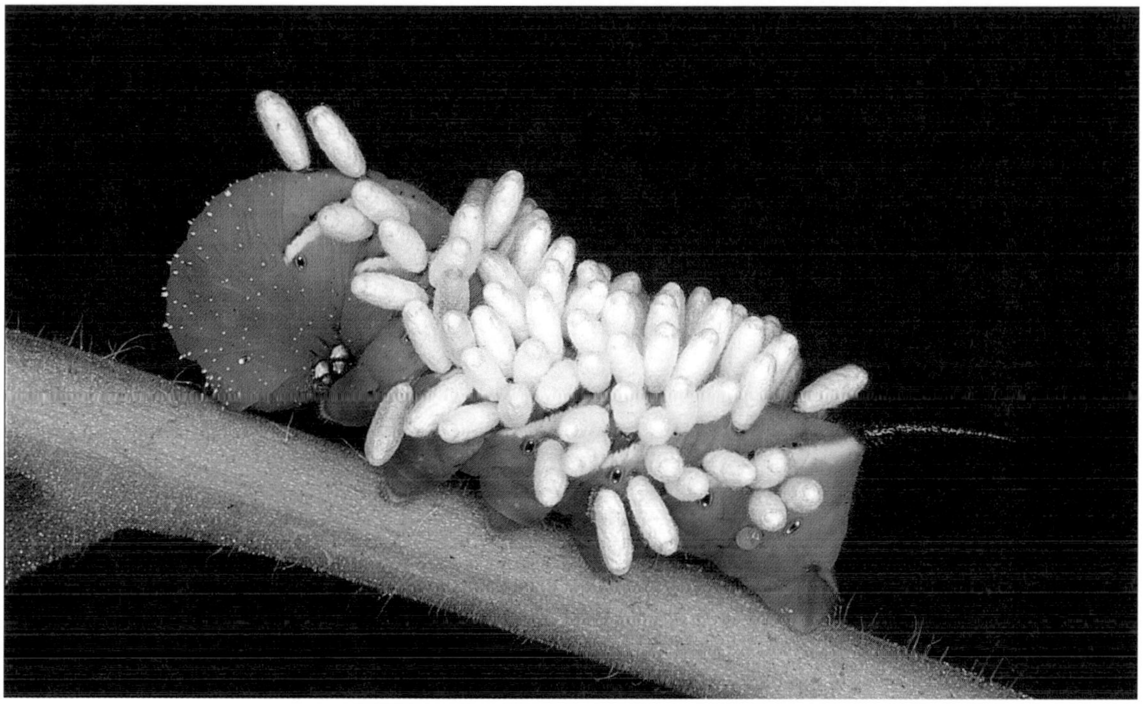

◆ Some parasitic wasps lay their eggs on the back of a tobacco hornworm. The larvae will emerge from the eggs and eat the hornworm.

Integrated Pest Management (IPM)

A system of PEST CONTROL in which multiple methods are used instead of synthetic chemical PESTICIDES. A major goal of Integrated Pest Management (IPM) is to reduce harm to the ENVIRONMENT caused by pesticides. However, synthetic pesticides may be used in combination with other pest-control methods. IPM reduces the number of pests to tolerable levels, but usually does not eliminate them entirely.

HOW IPM WORKS

Before IPM methods can be used, the relationships among the SPECIES in the pest's ECOSYSTEM must be investigated. For example, which PREDATORS eat the pest species? Which disease organisms cause harm to the pest species? How do other species compete with the pest species for food or space? After such questions are answered, a plan for integrated pest management can be developed.

One example of IPM is the case of the spotted alfalfa aphid in California. To combat the aphids, crop varieties that were naturally resistant to the INSECTS were planted. Then an INSECTICIDE was applied. Since resistant crop varieties were planted, less insecticide was necessary.

A common component of IPM is BIOLOGICAL CONTROL of pests, but others could also harm people and WILDLIFE, especially when they get into water sources. In addition, some pesticides remain in the environment long after their initial use. The toxic effects are passed up FOOD CHAINS as animals eat other organisms that contain the pesticide. IPM strives to keep pests under control without causing harm to the environment.

International Atomic Energy Agency (IAEA)

Agency of the United Nations that was founded in 1957 to promote safe and peaceful use of nuclear energy in its member nations. More than 100 nations are members of the International Atomic Energy Agency (IAEA).

The IAEA advises and assists its member nations in the development and use of nuclear materials in business, agriculture, medicine, and other nonmilitary fields. The agency also instructs world communities in how to build NUCLEAR POWER plants that meet the worldwide safety standards set by IAEA.

The IAEA is headquartered in Vienna, Austria. The agency publishes scientific reports about uses of nuclear energy. The agency is also responsible for research labs in Austria and Monaco, and the International Center for Theoretical Physics in Trieste—a city that is part of an Italian region established in 1947 as a free territory under the United Nations.

One job of the IAEA is to make sure that member nations conform to regulations in the United Nations Treaty on the Non-Proliferation of Nuclear Weapons. This treaty forbids the making of NUCLEAR WEAPONS. To carry out this task, the IAEA makes yearly inspections of member nations' nuclear facilities to account for all nuclear materials brought into or produced by that nation. IAEA assures that member nations also abide by other agreements, such as honoring the nuclear-weapon-free zones in the South Pacific and Latin America. [*See also* NUCLEAR FISSION; NUCLEAR FUSION; NUCLEAR WINTER; RADIATION; RADIATION EXPOSURE; RADIOACTIVE WASTE; and RADIOACTIVITY.]

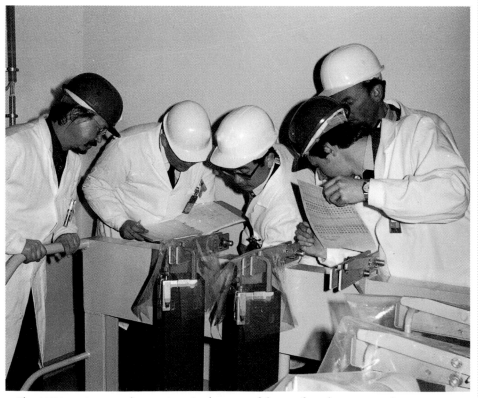

◆ The IAEA assists member nations in the peaceful use of nuclear materials.

International Convention for the Regulation of Whaling (ICRW)

▮▮International agreement among nations to limit HUNTING OF WHALES. Whales are marine MAMMALS that live throughout the world's OCEANS. One SPECIES, the blue whale, is the largest animal on Earth. Blue whales can reach lengths of up to 90 feet (27 meters) and weigh as much as 300,000 pounds (135,000 kilograms). Unfortunately, this species, along with the humpback, fin, and right whales, have been hunted almost to the point of EXTINCTION. The smallest members of the whale family include minke whales, dolphins, and porpoises.

Whaling is an important industry in some parts of the world. These animals are hunted for their meat and for the valuable oils they contain. Whale oil is used in many products, including perfume, soap, cosmetics, candles, and cooking oil. To preserve the various whale species, the International Convention for the Regulation of Whaling (ICRW) agreement was made in 1946. It was the first attempt by the nations of the world to protect whales in international waters.

INTERNATIONAL WHALING COMMISSION

In 1949, the INTERNATIONAL WHALING COMMISSION (IWC) was established to create and enforce all whaling laws.

◆ In 1994, a permanent whale sanctuary was created in Antarctica, the largest whale feeding ground in the world. Norway and Japan, who have the largest whaling industries, oppose the sanctuary.

The first decision by the IWC was to set a limit on the number of whales that could be caught in each year. The IWC also voted that people could hunt whales only within their nation's own territorial waters. However, throughout its history, the IWC has been criticized for its lack of effectiveness at enforcing these laws. Some countries, including France, Norway, Iceland, and Japan, protested and decided not to follow the IWC's policies. In addition, because no laws were passed to limit the type of whales that could be caught, many whale species with already low populations were overhunted to the brink of extinction.

SAVE THE WHALES

In the 1970s, there was public pressure on the IWC nations to "save the whales" by creating new laws to protect whales. In 1972, the IWC responded by starting a program in which member nations carefully monitored the whaling vessels of other members to make sure all laws were being followed. A few years later, in 1977, the IWC also passed a law urging nations to stop importing whale products.

In 1982, the IWC enacted a total ban on whaling to be in effect from 1986 to 1992. During this six-year period, the IWC would study whale populations throughout the world. Afterward, whaling would be allowed to resume only in certain areas for particular species. Norway, Japan, and the former Soviet Union elected not to be part of this agreement. The IWC allowed those countries to continue whaling for scientific purposes during the ban.

In 1992, when whaling was supposed to have resumed, many countries, including the United States, agreed to continue the ban. Other countries, such as Norway, Japan, and Iceland, disagreed. They argued that the ban would interfere with their traditional customs and eating habits.

Today, the whaling controversy continues. In 1994, a permanent sanctuary for whales was created in ANTARCTICA, the largest feeding ground for whales in the world. Norway and Japan, with the world's largest whaling industries, oppose the sanctuary. However, even if the current worldwide ban on whaling is lifted, commercial whaling in the whale sanctuary will be prohibited. [*See also* BIODIVERSITY; DOLPHINS/PORPOISES; ENDANGERED SPECIES; ENDANGERED SPECIES ACT; MARINE MAMMAL PROTECTION ACT; and VERTEBRATES.]

International Union for the Conservation of Nature and Natural Resources (IUCN)

▌▶ World organization consisting of governmental agencies, conservation organizations, and industry groups who work together to conserve WILDLIFE, solve global environmental problems, and advise governments on the best ways to manage NATURAL RESOURCES. Current membership in the International Union for the Conservation of Nature and Natural Resources (IUCN) includes nearly 200 governments and government agencies, and over 300 private conservation groups.

OBJECTIVES OF IUCN

Founded in 1948, IUCN has three basic conservation objectives:

1. to protect Earth's BIODIVERSITY as a valuable natural resource;

2. to make sure that Earth's natural resources are used wisely;

3. to educate humans on ways to live in harmony with other components of the BIOSPHERE.

To help meet its objectives, IUCN maintains an international network of over 5,000 volunteer scientists and wildlife professionals who serve on various committees dealing with the main aspects of IUCN's work. These include committees for ECOLOGY, Education, Environmental Planning, Environmental Law, National Parks and Protected Areas, and Species Survival. IUCN's headquarters are in Switzerland.

PUBLICATIONS

IUCN is probably best known for its work in protecting Earth's threatened and ENDANGERED SPECIES. IUCN publishes a comprehensive series of reference sources known as the *Red Data Books*. These books describe the current status of rare and endangered wildlife. Each book in the series provides information about a species' HABITAT, ecology, distribution, and major threats. Some titles in the series include: *IUCN Red List of Threatened Animals; Threatened Primates of Africa; Threatened Birds of the Americas;* and *Dolphins, Porpoises, and Whales of the World*. There are also titles dealing with threatened and endangered PLANT species, as well as a series that details action plans for CONSERVATION in effect for threatened species. [*See also* CONVENTION ON INTERNATIONAL TRADE IN ENDANGERED SPECIES OF WILD FAUNA AND FLORA (CITES); ENDANGERED SPECIES; ENVIRONMENTAL EDUCATION; ENVIRONMENTAL ETHICS; LAW, ENVIRONMENTAL; and SUSTAINABLE DEVELOPMENT.]

International Whaling Commission (IWC)

▶A worldwide organization set up to regulate whale HUNTING and protect WHALES from EXTINCTION. The International Whaling Commission (IWC) was formed as a result of the 1946 INTERNATIONAL CONVENTION FOR THE REGULATION OF WHALING (ICRW). The purpose of IWC was to prohibit the killing of right, bowhead, and gray whales, to establish whaling boundaries, and to set limits on hunting seasons for other whale SPECIES. The main function of IWC remains the setting of quotas, or limits, on the number of whales that can be killed. These limits are meant to protect whales from overhunting, while allowing the whaling industry to survive.

As environmental concerns grew in the 1960s and 1970s, IWC extended its protection policy to include the blue and humpback whales, while lowering quotas for other species. In 1985, IWC called for a temporary halt, or moratorium, on all commercial whaling. Although the halt is still in effect, IWC lacks the power to enforce its policies. Several member nations, including Norway and Japan, have continued commercial whaling.

Recently, both whaling and nonwhaling member nations have begun to object to the IWC quotas. The majority of IWC members believe all whales should be protected and are against whaling of any kind. Whaling nations want-

ing to protect their traditional industry would like to see all quotas removed. [*See also* ENDANGERED SPECIES; ENDANGERED SPECIES ACT; LAW, ENVIRONMENTAL; and MARINE MAMMAL PROTECTION ACT.]

Intertidal Zone

▶The region of a shoreline that lies between the highest and lowest levels of the tides. The coastlines of the world differ greatly from one region to another. Some are steep, while others are flat; some are smooth, while others are rough. Some coastlines are pounded by forceful waves; other coastlines are characterized by calm seas. In some areas very high tides break on the shoreline, while other shores have almost no tides at all. Despite these variations, each coastline has an intertidal zone—a

◆ California mussels inhabit rocks and become visible at low tide.

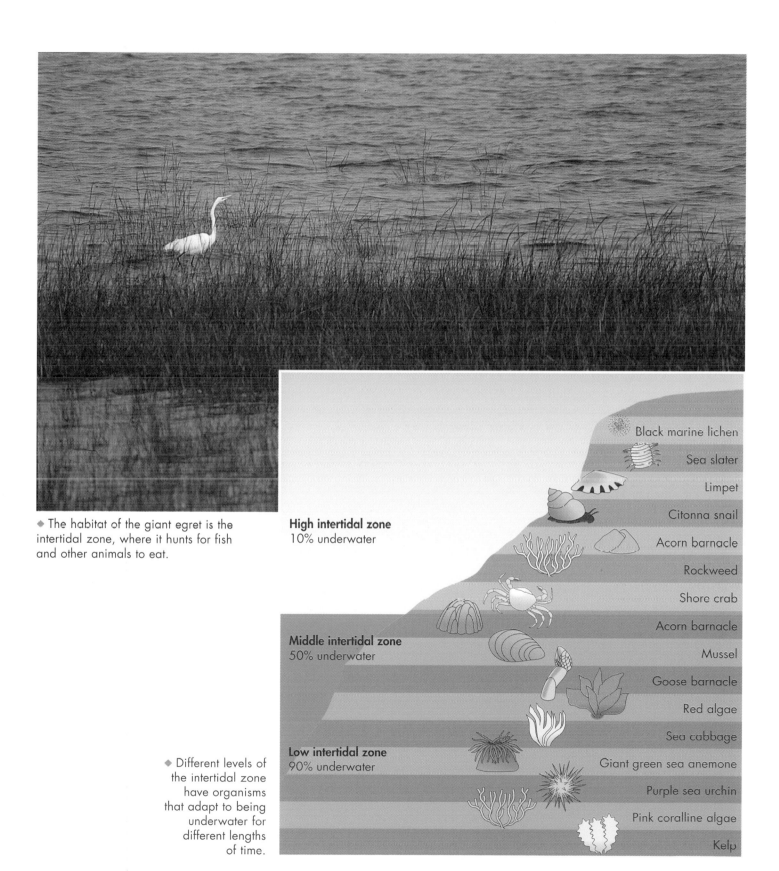

◆ The habitat of the giant egret is the intertidal zone, where it hunts for fish and other animals to eat.

◆ Different levels of the intertidal zone have organisms that adapt to being underwater for different lengths of time.

High intertidal zone
10% underwater

Middle intertidal zone
50% underwater

Low intertidal zone
90% underwater

Black marine lichen

Sea slater

Limpet

Citonna snail

Acorn barnacle

Rockweed

Shore crab

Acorn barnacle

Mussel

Goose barnacle

Red algae

Sea cabbage

Giant green sea anemone

Purple sea urchin

Pink coralline algae

Kelp

Intertidal Zone

region where the waves and tides wash up on the shore.

The intertidal zone is partly a marine ENVIRONMENT and partly a land environment. However, the features of the intertidal zone differ from all other environments. Waves and tides cover much of the intertidal zone for part of each day. At other times, the zone is exposed to the air and terrestrial WEATHER conditions. Waves, sand, and rocks may constantly tumble into the intertidal zone. In some latitudes, the shore may also be scoured by ice. Thus, organisms living in the intertidal zone must be able to survive both in and out of the OCEAN.

The intertidal zone is not uniform. Not only do tides come and go twice a day, but they rise and fall to different levels depending on the phase of the moon and on weather conditions. The lower edge of the intertidal region is always under water, except during the lowest tides of the month. The upper, or most inland, part of the intertidal zone is submerged by water only during the highest tides of the month or during severe storms. Thus, an organism at the bottom of the intertidal zone is usually under water, is tossed by waves, and lives where temperature and SALINITY, or saltiness, are fairly constant. In contrast, an organism at the top of the intertidal zone is more frequently exposed to air, direct sunlight, and extreme changes in temperature and salinity. As a result, different types of organisms survive best at different levels in the intertidal environment. This variety exists whether the shore is composed of sand, mud, swamp, or SALT MARSH, but it is especially easy to observe on a rocky shore.

A series of plant and animal combinations appears as horizontal stripes or belts, from the bottom to the top of rocky intertidal zones. This arrangement of organisms is known as *intertidal zonation.*

In general, communities in the intertidal zone can be divided into four main levels. At the bottom is the low intertidal zone. This level is usually submerged. Here, several kinds of communities may be present. In the tropics, there may be coral. In more northern areas, there may be a rich mat of brown ALGAE living with several other marine SPECIES. The middle intertidal zone, which is covered and uncovered by tides almost every day, is often crowded with various seaweeds, mussels, and barnacles. The barnacles are sessile, or nonmoving, organisms that live attached to rocks. In the upper intertidal zone, which is covered less often by tides, barnacles are also common. These animals close their shells tightly when exposed to air. When the tide comes in, they open their shells and filter the water for food. Predatory

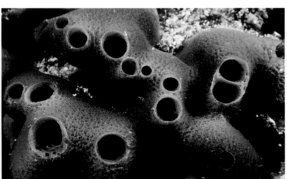

◆ In this colony, sponges (Porifera) take in water with small, edible pieces of decomposing material that they can filter out and eat. The water enters through pores all over their bodies and leaves through one big hole at the top.

◆ The intertidal zone of a rocky shore is home to a different community of plants and animals than that inhabiting a sandy shore. This is part of the Acadia National Park, Maine.

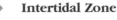

snails, which feed upon barnacles and mussels, are also common in the middle and upper intertidal zones. At the top of the intertidal region is the "splash zone." This region is dominated by snail species that feed on algae-covered rocks and LICHENS.

Although there are factors common to almost all intertidal zones, these areas are quite variable. Each has its own tides, temperature changes, and other physical conditions. Each also has its own collection of PLANTS, HERBIVORES, PREDATORS, and competing SPECIES. Because the most limiting factor for living things in rocky intertidal zones is space to attach to a rock, many species are competing for every square inch of space. This makes the intertidal zone one of the best places for studying COMPETITION. [*See also* BIOLOGICAL COMMUNITY; ECOSYSTEM; FOOD CHAIN; FOOD WEB; MARINE POLLUTION; NATIONAL SEASHORE; NATURAL DISASTERS; and OCEAN CURRENTS.]

◆ For protection, clams burrow under mud. They have siphons, like straws, that take in water from above and filter out small organisms and pieces of organisms to eat.

◆ Cleaner shrimp keep the tentacles of anemones free of uneaten bits of food.

Introduced Species

See EXOTIC SPECIES

Invertebrate

▶A general term applied to a multicellular animal without a vertebral column, or backbone. Invertebrates have distinct body shapes and sizes and may or may not have an external body covering such as a shell or exoskeleton, as adults. Invertebrates may have either radial symmetry, in which the body parts are arranged around a central point, or bilateral symmetry, in which one half of the body mirrors the other half.

Among the many different groups, or **phyla**, of invertebrates are Porifera (sponges), Cnidaria (coelenterates such as jellyfish, hydrozoans, sea anemones, and corals), Ctenophora (comb jellies), Platyhelminthes (worms with flat bodies such as tapeworms), Nematoda (roundworms), Annelida (segmented worms), Phoronida (horseshoe worms), Brachiopoda (lamp shells), Bryozoa (moss-animals), Mollusca (soft-bodied animals that usually have limy shells such as

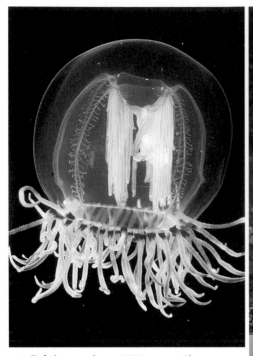

◆ Jellyfish are about 90% water. These fragile-looking Cnidarians are all carnivores.

◆ The freshwater counterpart of the spiny lobster is the crayfish, which is a smaller arthropod.

snails, clams, squids), Arthropoda (insects, arachnids, crustaceans), and Echinodermata (animals with external spines such as sea stars, brittle stars, sea urchins, sea cucumbers, sea lilies). Invertebrates comprise about 97% of the 1.7 million animal organisms that have been identified and given scientific names.

Borehole tracks preserved in rock suggest that invertebrate marine animals arose from one-celled **protozoans** over 1,000 million years ago. Jellyfishes, worms, and other **metazoans** appeared 700 million years ago. The great age of invertebrate development was the Paleozoic Era, which lasted from 570 million to 225 million years ago. During this era, most of the major phyla appeared and gave rise to VERTEBRATES, animals with backbones. By 400 million years ago, invertebrates such as scorpions and millipedes had established themselves on land. Most invertebrate groups known from that time still thrive.

The phylum Arthropoda includes 1,300,000 known SPECIES of insects, more than the number of species in all other animal phyla combined. This phylum is made up of animals that have an outer skeleton of armorlike chitin and have jointed legs. Estimates of the total number of known and unknown insect species range from 4,400,000 to 50 million—30 million of them in tropical RAIN FORESTS.

INSECTS and other invertebrates play vital roles in the structure and functioning of ECOSYSTEMS. Both as pollinators and eaters of seed, invertebrates influence plant-species composition in regions throughout the world. Burrowing animals, especially earthworms, improve the texture and **aeration** of SOIL and make MINERALS more available to plant roots. As food, invertebrates sustain large populations, human as well as animal. For example, in a marine FOOD CHAIN, tiny invertebrates such as copepods graze on the even tinier organisms that form PLANKTON. Copepods, in turn, are eaten by FISH that are food for bigger fish that may be eaten by seals, WHALES, or human beings. [*See also* BIODIVERSITY; BIOLOGICAL COMMUNITY; FLOWERING PLANT; POLLINATION; SPECIES DIVERSITY; and WILSON, EDWARD OSBORNE.]

Ionosphere

See ATMOSPHERE.

Irrigation

◗ The watering of land by artificial methods. Irrigation provides a method for watering crop PLANTS. Although used in many areas, irrigation is most often used in regions that receive little rainfall. For example, irrigation is used in some DESERT regions. Irrigation is also common in areas that experience long periods of drought. About 70% of all fresh water used by humans today is for irrigation.

◆ In flood irrigation, an entire area is covered with water, as in the walnut orchard shown here.

The importance of transporting and directing water to where it is needed has been known to humans for thousands of years. Archaeological evidence from Egypt indicates that people used irrigation for farming as early as 3000 B.C. Ancient Egyptians built large canal systems that carried water from the Nile River to their crop fields. Similar methods of irrigation were used in China and India around the same time. The Aztec and Inca civilizations of Mexico and South America were also based on agriculture that used irragation.

Modern irrigation methods in the United States began in the 1840s. At this time, settlers irrigated crops in the Salt Lake Valley of Utah. Today, the use of irrigation has spread to nearly every part of the country. However, most irrigation occurs in California and hot, dry states of the southwestern United States. The BUREAU OF RECLAMATION under the DEPARTMENT OF THE INTERIOR of the United States provides much of the planning and funding for major irrigation projects.

SOURCES OF IRRIGATION WATER

Crops and other plants require fresh water for growth. The two main sources of fresh water on Earth are groundwater and SURFACE WATER. As its name suggests, groundwater is water that soaks into SOIL. Much of Earth's groundwater is stored in AQUIFERS. To recover water from an aquifer, an irrigation well is dug at or near the area where the water is needed. Water is then pumped into a ditch or a system of pipes that carry it to the crops.

Surface water is the water in lakes, ponds, rivers, and streams. Most water used for irrigation comes from surface water. In many countries, DAMS built across rivers and streams create large RESERVOIRS for irrigation water. In some areas, farmers tap into surface water supplies by building long networks of canals from local streams to their land. The canals divert water from the streams to the crops. If a farm is at a lower elevation than its water source, water is carried to the needed area by gravity. For example, a farm in a valley may receive the meltwater from snow on a nearby mountain. In contrast, when a plot of land is higher than its water source, pumps must be used to deliver water to the crops.

PROBLEMS OF IRRIGATION

Irrigation is not without problems. A major problem faced by farmers using irrigation is that much water is lost before it reaches crops. For example, in hot, dry areas, large amounts of irrigation water are lost to the air through evaporation. In regions with porous soils, water may seep into the ground before reaching crops. Seepage is sometimes controlled by lining reservoirs and irrigation channels with asphalt

or concrete. Evaporation is a more difficult problem to prevent. However, one method used to reduce evaporation is the building of smaller area, but deeper, reservoirs. This technique reduces the evaporation rate from the reservoir.

TYPES OF IRRIGATION SYSTEMS

The proper use of irrigation requires skill on the part of the farmer. To run a cost-effective farm, farmers must know when to irrigate their crops, how much water to use, and what method of irrigation is most suited to their land and their crops.

Four irrigation methods are available to farmers. The decision about which method to use depends on a variety of factors. These factors include the distance of the crops from the water source, the ability of the soil to hold water, and the type of crop being irrigated.

Surface Irrigation

The most common method of irrigation is *surface irrigation.* In surface irrigation, water is spread over the surface of the field. *Flood irrigation* is one type of surface irrigation. Flood irrigation involves covering an entire field of crops with water. Small soil walls are sometimes built to hold the water in place on the field. This method of irrigation works best on flat, level crop fields on which plants that require much water are grown. Alfalfa and rice are the two main crops grown with the use of flood irrigation.

Another type of surface irrigation is called *furrow irrigation.* In furrow irrigation, narrow ditches called *furrows* carry water between rows of crops. Such irrigation is used on fairly flat ground for crops that need moderate amounts of water. Corn, cotton, and potatoes are crops that are irrigated in this way.

Drip Irrigation

Drip irrigation was developed in Israel in the 1950s because of the problems associated with evaporation. Drip irrigation conserves water by applying it directly to crops in small amounts. In this method, water is carried through plastic tubes that lie on or under the ground. Small openings in the tubes allow water to trickle directly into the soil near the root systems of plants. Drip irrigation can be used on all types of crops and on different soils. The method is ideal for areas that have small supplies of irrigation water. However, the high cost of installing the underground pipes makes it impractical for most farmers. Today, movable plastic above-ground pipes are increasingly being used.

Sprinkler Irrigation

In sprinkler irrigation, water is carried to a field of crops by long pipes. The water is then pumped from the pipes and sprayed onto plants through sprinkler heads. Sprinkler systems, such as those often used on lawns and gardens, distribute water over a large area. However, such distribution wastes

◆ Sprinkler irrigation is supplied to an alfalfa field in the Palouse River Valley in Washington State.

◆ In deep irrigation, water is applied directly to crops in small amounts through plastic tubes lying on or under the ground.

◆ A California bean field is being irrigated by syphons that carry water to the furrows.

a great deal of water by spraying it over areas where it is not needed. In addition, water droplets sprayed into the air are easily carried to other areas by wind. Most sprinkler systems used on farms are moveable. These self-propelled systems move across a crop field without requiring attention by a farmer. Sprinkler irrigation can be used for most crops and is ideal on most land surfaces. [*See also* AGRICULTURAL REVOLUTION; AGROECOLOGY; DUST BOWL; GREEN REVOLUTION; WATERSHED; and WATER TABLE.]

Ivory-billed Woodpecker

❙❭Critically endangered North American SPECIES of woodpecker. The largest of the North American woodpeckers, this species can reach up to 20 inches (50 centimeters) in length. Males are characterized by a prominent red crest on the head. Females possess a black crest.

Once common in the southeastern United States, the ivory-billed woodpecker has been driven to the brink of EXTINCTION by logging and overhunting. The decline of the woodpecker began in the late nine-teenth century, when the timber industry cut down most of the woodpecker's HABITAT. There have been some unconfirmed sightings of the ivory-bill, but most scientists believe that the species has been eliminated in the United States. The few remaining individuals of the species are living in a small patch of forest in Cuba. [*See also* ENDANGERED SPECIES; EXTINCTION; and WILDLIFE MANAGEMENT.]

◆ The ivory-billed woodpecker is considered by most authorities to be extinct in the United States.

K

Keystone Species

▷ A group of organisms that are especially important in determining the structure of a BIOLOGICAL COMMUNITY. A keystone is the stone at the top of an arch that cannot be removed without causing the arch to collapse. The term *keystone species* refers to SPECIES that cannot be removed from a community without causing radical change in it. The change could take several forms: the disappearance of other species, a shift to a different type of plant cover, or even a switch between a wet HABITAT and a dry one.

Keystone species are not always the same as the dominant species of a community. A dominant species is one that is abundant, but its disappearance might not disrupt the community very much. Keystone species may be less abundant, but more important to the community as a whole. Often they are important because they create habitat or suppress species that would otherwise dominate the community. Both these effects can increase SPECIES DIVERSITY.

TYPICAL KEYSTONE SPECIES

Some of the best examples of keystone species come from aquatic ENVIRONMENTS. For example, Pisaster, a large predatory starfish, is the key to maintaining the diverse community in the rocky INTERTIDAL ZONE on the West Coast of North America. Pisaster preys mainly on mussels. Where Pisaster was removed by a researcher, mussels became much more abundant. Mussels competed with other organisms for space, finally covering the rocks in such numbers that most barnacles, limpets, and other species were excluded, and the rocky intertidal community was greatly altered. In a similar way, the great **kelp** FORESTS of the West Coast depend in large part on the presence of the sea otter, a PREDATOR of sea urchins. If sea urchins become too abundant, they overgraze kelp beds and destroy them. This removes the habitat needed by many FISH and other marine species. By eating sea urchins and controlling urchin populations, sea otters help to maintain the kelp forests. In kelp beds off the East Coast of North America, there are no sea otters, but sea urchin populations seem to be controlled by a different keystone species, the lobster.

On land, some of the best known keystone species are large MAMMALS. The African ELEPHANT, for example, creates GRASSLAND communities where there would otherwise be woodland communities. It

---

THE LANGUAGE OF THE ENVIRONMENT

kelp a species of large brown marine alga in the genus *Laminaria*. Some kelps are very large—up to 135 feet (40 meters) long—and grow in dense stands called *kelp beds* or *forests*. These forests are very productive and support a large community of marine organisms.

prairie dogs small burrowing rodents found in North America.

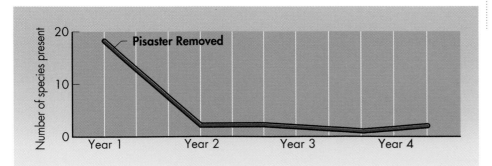

◆ Pisaster disaster: This graph shows what happened when a keystone species, the Pisaster starfish, was removed from its biological community. The number of other species immediately began to decline.

does so by feeding on shrubs and trees, often killing them in the process. Reducing the number of trees encourages the spread of grasses. Grass fires then help prevent trees from growing back, resulting in a grassland community that supports a variety of species, from INSECTS to lions and antelope.

SEARCHING FOR KEYSTONES

Most keystone species are less obvious than elephants and sea otters. In fact, it can be quite difficult to identify a keystone species in a community, except by removing it and watching the community change into something else. The role of Pisaster was discovered by deliberate experiment. The need for sea otters and lobsters in kelp forests was shown when these animals were overharvested, and the kelp disappeared. Removing species to observe the consequences is often impractical or harmful, so relatively few keystone species are known. It is not even known whether they are common or unusual. There are, after all, many ways to explain changes in communities without assuming that keystone species are involved.

KEYS TO CONSERVING COMMUNITIES

It is important to recognize keystone species wherever they do exist, because conserving them benefits many species at once. Saving one species at a time can be less efficient, as in the case of black-footed ferrets on the North American PRAIRIE. Restoring this ENDANGERED SPECIES by breeding it in captivity has been a long, expensive project. A different approach to saving ferrets would be to save **prairie dogs**, which are keystone species of the prairie community. Prairie dog "towns" not only provide food and habitat for black-footed ferrets but support a wide variety of other species as well. One recent study identified about 170 VERTEBRATE animals that depend in

◆ American alligators are a keystone species in the Everglades. "Gator holes" dug by these reptiles provide water and habitat for many species during the dry season.

some way on the presence of prairie dogs. Despite this, prairie dogs are still routinely exterminated.

A more encouraging example is that of the beaver, which is a keystone species for pond communities. Beavers in North America have now returned to areas where they were once exterminated. By damming streams, creating ponds, and cutting trees along pond edges, beavers provide habitats for a wide variety of turtles, AMPHIBIANS, fish, other aquatic animals, BIRDS, and wetland plants. This helps to maintain BIODIVERSITY and thus slow down HABITAT LOSS. If such keystone species can be identified and wisely managed, then it may become easier to conserve biological communities rather than individual species. [See also WILDLIFE CONSERVATION and WILDLIFE MANAGEMENT.]

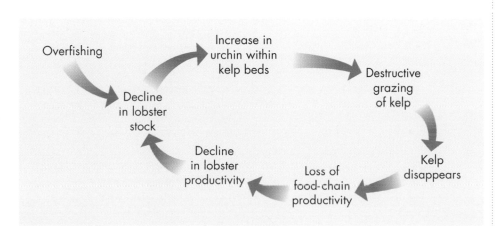

◆ The link between overfishing of lobsters and destruction of kelp beds by sea urchins is illustrated above. Lobsters are not the only influence on the sea urchin population, but they probably are important in controlling their number.

L

Labeling, Environmental

▐▌ Product labeling intended to persuade potential consumers that using the product is somehow beneficial, or not destructive in any way, to the ENVIRONMENT. Private organizations in the United States and governments of several countries have all attempted to develop environmental labeling. Notable examples are Germany's Blue Angel (the earliest such label, launched in 1978), Japan's Eco-Mark, and Canada's Environmental Choice.

Environmental labeling makes frequent use of terms that have legal meaning, such as "recyclable" and "made from recycled material," and of some terms that do not, such as "natural" and "earth-friendly." It is a rare product, however, that truly benefits the environment, and ultimately, one that has not been fashioned from raw materials or extracted from it. Thus, while some uses of environmental labeling—such as on products made from recycled materials—may be legitimate, others can be regarded only as cynical attempts to manipulate consumers who are concerned but also confused about environmental matters. An example is the supposedly BIODEGRADABLE plastic trash bag. One type is made of a plastic that becomes brittle and breaks apart after prolonged exposure to sunlight. However, almost all trash bags of every type end up where sunlight is unable to reach them—buried in LANDFILLS. Another type of trash bag, made of a blend of ordinary plastic, cornstarch, and a special chemical, is designed to degrade when buried. Soil bacteria eat the cornstarch, weakening the bag, and the special chemical works on the plastic so that the bag eventually falls to pieces. However, these pieces remain in the environment.

Difficult questions arise from efforts to make products deserving of environmental labeling. For instance, which is environmentally preferable, an energy-efficient appliance that often breaks down and has to be transported in a vehicle to a repair shop or a less efficient appliance that never breaks down? Consumers trying to determine whether products are environmentally "good" or "bad" may discover that it is not always easy. [*See also* RECYCLING, REDUCING, REUSING.]

◆ The label on this package of napkins indicates that it is made from recycled paper.

◆ Some product labels carry the symbol to indicate that the container is recyclable, such as the juice bottle shown above.

Landfill

❿ A site where SOLID WASTES (GAR-BAGE) are disposed of by burial in the ground. The average person in the United States produces more than 1,800 pounds (810 kilograms) of garbage each year. Today much of this garbage, especially news-papers, aluminum cans, and glass jars and bottles, may be recycled or reprocessed to make new products. Some of the garbage may also be incinerated, or burned. The remain-ing solid wastes are often buried in the ground in places called *land-fills*.

◆ Landfills are built so people will dump their garbage in them instead of by the roadside.

THE HAZARDS OF DUMPING

The landfill disposal method was developed in England in 1912. How-ever, this method of garbage dis-posal was not used in the United States until the 1960s. Before land-fills were used in the United States and other countries, garbage and other solid wastes were often placed in open dumps on land and thrown into **waterways**.

The use of open dumps and OCEAN DUMPING as methods of garbage disposal created health, **aesthetic**, and environmental prob-lems. For example, garbage dumps on land provided excellent breed-ing grounds for mice, rats, and various unwanted INSECTS and microorganisms. Many of these organisms carried and spread dis-ease. The piling of garbage, as well as the activity of rodents, is an unwelcome sight to most people. In

THE LANGUAGE OF THE ENVIRONMENT

aesthetic something of beauty.

waterways bodies of water deep enough for boats and ships.

addition, as garbage in dumps breaks down through the natural process of DECOMPOSITION and WEATHERING, foul odors are often given off. These odors not only per-

meate the dump site but annoy sur-rounding communities as well. Open dumps also create environ-mental hazards in the form of AIR POLLUTION, land pollution, and WATER POLLUTION.

Ocean dumping and the dis-posal of waste in rivers create many of the same problems as open dumping on the land. Often these problems are worsened as flowing water transports the garbage. For example, garbage thrown into water in one area may be carried to the shores of a neighboring com-munity or other areas. Water also provides a suitable HABITAT for many disease-causing organisms. As wastes are dumped into water, these organisms may be provided with nutrients that help them thrive. In 1892, New York City suffered from severe outbreaks of cholera and typhoid. Both these diseases were spread by microorganisms living in garbage-infested waters.

THE HAZARDS OF LANDFILLS

Landfills were created to solve some of the problems that resulted from open dumping on land and in waterways. The first landfills were made by dumping garbage into natural or human-made depressions, or holes, in the ground. Each day, garbage trucks, carrying all kinds of wastes, dumped their loads into the depression. At the end of the day, bulldozers covered the new landfill contents with SOIL. This process helped reduce the odor created by the garbage. In addition, landfills supported fewer unwanted organisms, such as rodents and insects, than did open dumps.

While these early landfills seemed to be an improvement over open dumps, they too created environmental hazards. The landfills were unattractive to look at and still gave off unpleasant odors that spread to communities surrounding the landfill. Although landfills had been in use for only a short time, by the 1970s environmental groups discovered that hazardous gases, including METHANE, seep into the air from decaying garbage. In California, landfill gases were discovered in nearby buildings.

Water pollution is a serious result of landfills. Rainwater washing through a landfill picks up and carries chemicals contained in the decaying garbage and soil. This water, called *leachate*, may transport the chemicals into groundwater or nearby lakes, OCEANS, or streams. Groundwater often flows in underground rivers called AQUIFERS. In many parts of the United States, aquifers, like lakes and

◆ Garbage is brought to this sanitary landfill, where it is compacted and covered with soil.

rivers, are a major supply of drinking water. Such drinking supplies are often contaminated by leachate-carrying toxic chemicals.

SANITARY LANDFILLS

To solve some of the environmental problems associated with early landfills, many communities now make use of sanitary landfills. Wastes placed in sanitary landfills are buried in layers, usually about 10 feet (3 meters) deep. The waste layer is then compacted by bulldozers and covered by a soil layer about 6 inches (15 centimeters) deep. Once covered, the soil layer is again compacted by bulldozers. This process is repeated again and again until the landfill reaches a

predetermined height. Once that height is reached, the landfill is closed and covered over by another soil layer that is about 24 inches (60 centimeters) thick. Trees and grasses may then be planted on top of the landfill to improve its appearance and help prevent EROSION of the soil.

To avoid problems caused by LEACHING, sanitary landfills are lined with layers of PLASTIC or clay. These materials create a barrier between the landfill site and surrounding soil. The barrier is designed to prevent rainwater that enters the landfill from flowing out. In addition, drainpipes that capture water are often placed around the landfill in case any leaching occurs. The water that enters the drainpipes is

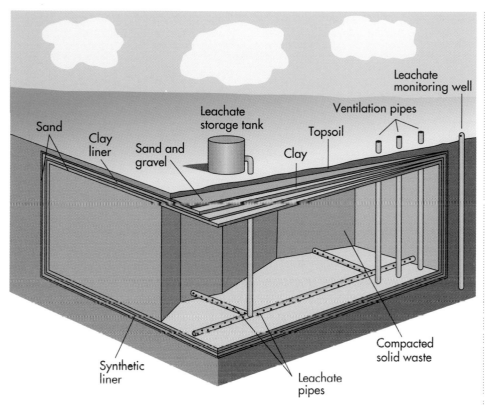

Sand

Clay liner

Sand and gravel

Leachate storage tank

Leachate monitoring well

Ventilation pipes

Topsoil

Clay

Compacted solid waste

Synthetic liner

Leachate pipes

◆ A well-constructed landfill has several linings of synthetic material, sand, or clay and has ventilation pipes to allow gas to escape. There is a leachate collection and monitoring system as well.

◆ Housing tracts are being built in these landfill terraces.

collected and sent to WATER TREAT-MENT plants for cleaning. To prevent the escape of toxic gases, ventilation pipes are placed in the ground. The ventilation pipes either control the release of gases into the ENVI-RONMENT or collect and store the gases. In some areas, the collected methane is sold for use as FUEL.

PROBLEMS OF SANITARY LANDFILLS

Sanitary landfills are a major improvement over earlier landfills and methods of waste disposal. However, such landfills also have drawbacks. One problem is their large size. In areas where populations are dense or land is needed for other purposes, such as farming, finding suitable sites for landfills may be difficult. POLLUTION of the water supply is another risk. Although precautions are taken to prevent such pollution, breaks or tears in liners may allow polluted water to leach out into neighboring soil or ground water supplies.

The construction and maintenance of landfills is also very costly. The price of the site for the landfill may be high. Excavation, hauling wastes to the site, and covering and closing it are expensive, too. In addition, landfills tend to subside, or sink, as the organic wastes they contain decompose. Therefore, additional landscaping may be needed long after the site has been abandoned. If the site has been reclaimed or developed for other uses, such as office complexes, sports arenas, or parks, subsiding soil may cause serious structural damage to buildings and equipment located there.

ALTERNATIVES TO LANDFILLS

Between 1982 and 1989, nearly 3,000 landfills in the United States were closed. By the year 2000, more than half the cities in the United States are expected to shut down their landfills. The two main reasons for the closing of landfills are environmental hazards and the NIMBY syndrome. Nimby is an acronym for "not in my backyard." While most people recognize the need to dispose of garbage somewhere, few want a landfill in their neighborhood. Such community opposition has closed or prevented landfills from opening in some areas.

Many communities are looking for alternatives to landfills while at the same time encouraging people to produce fewer solid wastes. To cut back on the amount of solid wastes entering landfills, a number of states and communities have passed recycling laws. These laws may require people to separate their garbage so that materials such as paper, plastic, ALUMINUM, steel, and glass go to recycling centers instead of local landfills. In addition, some communities forbid people to dispose of organic wastes in the form of lawn clippings and leaves with their garbage. In many areas, people are encouraged to mulch, or cut up into small pieces, these materials and use them as fertilizer for their lawns and gardens. COMPOSTING of lawn scraps, along with food scraps and other organic wastes, is another popular alternative to disposing of materials in landfills. Composting involves the natural breakdown of organic wastes by microorganisms. As these materials are broken down, soil rich in organic matter is produced.

In addition to producing less garbage, many communities are actively searching for other methods of disposal. These methods may include incineration, or the burning of garbage, and the creation of trash-to-steam plants. These plants burn garbage and then use the heat given off to create steam for generating ELECTRICITY. Until these and other WASTE MANAGEMENT techniques are developed, many areas will continue to use landfills as a method of disposal. [*See also* CONSERVATION; HAZARDOUS WASTE MANAGEMENT; SOLID WASTE DISPOSAL ACT; and SOLID WASTE INCINERATION.]

Land Stewardship

▶ The act of managing land wisely so that it is passed on to others in healthy condition. The word *steward* originally meant "manager of the house." A land steward is the manager of a portion of an ECOSYSTEM who tries to maintain or improve the health of that area.

A farmer who works to improve the quality of land for LIVESTOCK or crops would be considered a land steward. The scientists, **rangers**, and educators who oversee a NATIONAL PARK could be considered stewards of that land. A person who volunteers in a local **preserve**, or homeowners who decide to improve their yards for WILDLIFE are all examples of stewardship.

FARMERS AND FORESTERS AS STEWARDS

Farmers who try to improve the quality of their land are stewards. They try to ensure that their land will be fertile for generations to come. Working to improve the SOIL, and preventing EROSION, and using ORGANIC FARMING techniques and other methods are all ways in which the farmland can be left in better condition.

A FOREST can also be managed with stewardship in mind. Foresters might use a variety of methods of SILVICULTURE to encourage the growth of profitable trees, particular wildlife, or general BIODIVERSITY.

STEWARDS OF NATURAL AREAS

Many people are involved as stewards of natural areas, either professionally or as volunteers. They may be involved in PUBLIC LANDS or private lands owned by organizations such as The Nature Conservancy

---

THE LANGUAGE OF THE ENVIRONMENT

preserve an area restricted for the protection and preservation of natural resources.

rangers keepers of specified areas, such as forests.

A wise land steward is someone who tends the land and its resources to preserve them for future generations. An unwise steward abuses the land and its resources, making them unusable now and in the future.

(TNC). TNC is a nonprofit organization that acquires and manages important natural areas. Especially important to Conservancy stewards is the maintenance of biodiversity.

Stewardship of natural areas might involve the elimination of EXOTIC SPECIES and the encouragement of NATIVE SPECIES. A steward of a prairie preserve, for instance, might schedule a PRESCRIBED BURN at the right time of year to kill off exotic shrubs and wildflowers and promote the growth of native PLANTS. If the preserve is large enough, the same steward might release bison, since grazing them encourages the growth of native plants and animals.

STEWARDSHIP IN THE BACKYARD OR NEIGHBORHOOD

Stewardship does not have to occur only in wild areas or on farms. People can act as stewards of their own backyards by encouraging native plants, creating small ponds for wildlife such as AMPHIBIANS, or putting up nesting boxes for particular species of BIRDS. By contacting local nature centers, state Fish and Wildlife agencies, The Nature Conservancy, or the National Wildlife Federation, anyone can start to learn about being a steward of his or her own small part of the world.

◆ The stewardship of this land would be improved by laws banning the practice of strip mining or by enforcing laws mandating the return of strip-mined land to pre-mining condition.

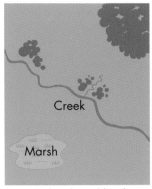

Creek

Marsh

Undeveloped land

Typical housing
development

Swimming pool
cluster

Creek

Cluster
housing

Pond

Cluster housing
development

◆ When deciding how available land
should be used, people must consider
not only their need for factories, farms,
jobs, and financial gain but also the
needs of the community at large,
including the plants and animals that
share the ecosystem.

Land Areas Administered by Federal and State Agencies		
Agency	**Landholding**	**Millions of Acres**
Bureau of Land Management	National Resource Lands	244
National Park Service	National Park System	79
U.S. Fish and Wildlife Service	National Wildlife Refuges	89
U.S. Forest Service	National Forests	191
State and local parks and forests		25

Land Use

❿ Ways in which people use land
and its NATURAL RESOURCES, including
not only the SOIL, water, and MIN-
ERALS but also any plant and animal
ECOSYSTEMS. Land may be used in
many different ways, such as for
farms, recreational parks, wildlife
refuges, mines, factories, housing
developments, highways, or cities
filled with skyscrapers. In choosing how to use land for their own
benefit, people often change the
ecosystem.

In the past, animals and PLANTS
in areas that were to be developed
were uprooted for the needs or
desires of humans. Now, more people use COST-BENEFIT ANALYSIS to
decide the worthiness of a new
project, looking not just at any
profit on money invested but at any
advantages or disadvantages to
people, animals, and plants in the
ecosystem. Before such projects are
begun in most areas of the United
States, the concerns of its citizens
are heard, either by inviting

◆ Many of Earth's natural resources
are used for recreational purposes,
such as the Kennebago River,
where fly fishing is popular.

◆ As cities grow, more pieces of land around them are developed for housing purposes.

citizens to speak for or against a proposed project in an open town meeting or by having citizens vote on possible projects at election time in a referendum.

Land-use decisions can be made on a small, local scale or can be debates that involve an entire nation. For instance, in some states, anyone wishing to construct a new building or create any significant changes on his or her land must appeal to a local conservation or wetlands commission. The commissions have the authority to approve, stop, or modify the planned work, based on its impact on the ENVIRONMENT. An example of a national land-use issue in the United States is the continuing debate on whether the ARCTIC NATIONAL WILDLIFE REFUGE (ANWR) is an appropriate place for oil exploration and OIL DRILLING. Land-use decisions often are a source of conflict between

Types of Land and Their Uses

Land Characteristics	Primary Uses
Cultivatable land	
1. Flat, with good drainage	Agriculture
2. Slight slope, sandy soil, or poor drainage	Pasture, contour farming, strip cropping
3. Moderate slope, limited soil or drainage	Pasture, contour or terrace farming, watershed
4. Severe slope, limited soil or drainage	Pasture, orchards, limited farming, urban industry
Noncultivatable land	
5. Steep slope or rocky, shallow, or sodden soil	Grazing, forestry watershed, no plowing
6. Moderately limited for grazing and forestry	Limited grazing and forestry, watershed, urban industry
7. Severely limited for grazing and forestry	Grazing, forestry, watershed, recreation, wildlife, urban industry
8. Steep slope, poor soil, too much or too little water	Recreation, watershed, wildlife, urban industry, wholly unsuitable for grazing and logging

groups that wish to protect a HABI-TAT in its present state and those who are in favor of the economic benefits that come from certain types of land use.

USING AND ABUSING LAND

How people use available land depends a great deal on the land's natural qualities and the needs of the people in the region. For example, one region might have abundant farmland and FORESTS but a small population; another region might have only a small amount of fertile land but a much larger population. In places with limited farmland, people often misuse the land to grow as much food as possible.

In some places, the quick clearing of trees for farming, GRAZING, and firewood destroys forests. This action has many consequences for the environment. It can leave the forest soil unprotected against water and wind EROSION. It destroys the habitat of many INSECTS and BIRDS. And when forests disappear, the CLIMATE can change because trees play an essential part in maintaining local patterns of Earth's temperature and rainfall.

Other examples of land misuse include overdevelopment, STRIP MINING without appropriate reconstruction of the land, OVERGRAZING, and disposal of TOXIC WASTE in areas near rivers and streams. With a growing human population, questions of land use will continue to be sources of controversy. [*See also* BIOLOGICAL COMMUNITY; DUST BOWL; EXTINCTION; ECOLOGY; ENVIRONMENT; ENVIRONMENTAL ETHICS; FORESTRY; and LAND STEWARDSHIP.]

Law, Environmental

▶ That body of law concerned with preventing damage to, or preserving, NATURAL RESOURCES and the ENVIRONMENT. Every country has enacted laws that are designed to prevent POLLUTION and depletion of natural resources. These laws cover a wide range of subjects, such as preserving HABITATS and BIODIVERSITY, controlling hazardous air pollutants, and cleaning up polluted rivers. Many environmental laws are international. Nations get together and agree to laws designed to protect resources that are important to everyone, such as WHALES in the OCEAN, or the OZONE LAYER in the ATMOSPHERE.

An environmental law often works in two ways—through punishment and through reward. For instance, under the provisions of the U.S. CLEAN WATER ACT, manufacturers who release air pollutants into the atmosphere can be fined; cities that take actions that the law is designed to encourage may likewise be provided with funds to upgrade their SEWAGE systems. The law states which government agency is responsible for enforcing the law.

THE HISTORY OF ENVIRONMENTAL LAW

For thousands of years, people have understood that laws are needed to protect the environment. Two thousand years ago, Julius Caesar banned iron-wheeled char-iots from the Roman cobblestoned streets at night in an attempt to control NOISE POLLUTION, a problem that has become even worse in modern industrial societies. Ancient China had laws against polluting water.

Most environmental laws, however, have been enacted in the last 200 years, since industrialization. Early in the nineteenth century, societies started to use large amounts of FOSSIL FUELS, which have the potential to pollute the environment on a large scale.

The twentieth century is notable for new laws designed to solve problems caused by the human population explosion. The worst of these problems is habitat destruction. As people have spread over the face of the globe, natural ECOSYSTEMS have been replaced by farms and towns. FORESTS have been cut down, leading to soil EROSION, CLIMATE CHANGE, and wood shortages in many countries. Ecosystems have been destroyed, leading to the EXTINCTION of many SPECIES and the resulting loss of many valuable products. The first major attempts to preserve habitats for future generations came at the end of the nineteenth century.

By 1872, the destruction of natural areas in the United States was so extensive that the government realized that some provision must be made for preserving PUBLIC LAND. President Ulysses S. Grant signed the act designating thousands of square miles of Wyoming as YELLOWSTONE NATIONAL PARK. Since then, many other states and other countries such as New Zealand and Costa Rica have also set aside land to be preserved.

HUMAN HEALTH

Environmental laws are often designed to protect health. In many countries, though not in the United States, a single government agency is responsible for both medical care and the environment, in the belief that pollution and other environmental problems are primarily threats to public health. Many diseases are caused, at least in part, by environmental conditions. For convenience, we can divide these diseases into several categories:

1. Diseases caused directly by pollution, such as RADIATION sickness and respiratory diseases caused by breathing polluted air.

2. Infectious diseases caused by organisms and VIRUSES, which are transmitted in polluted environments. These include the many diseases caused by organisms that live in polluted drinking water, such as cholera and hepatitis.

3. Infectious diseases caused or made worse by human beings living very close to one another or to other sources of infection. One of these is rabies, which is transmitted to humans by the bite of an animal infected with the disease. Others are Lyme disease, malaria, and hantavirus.

THE DIFFICULTIES OF LEGISLATION

Environmental laws are necessary because protecting the environment requires effort and, sometimes, the expenditure of money. For example, it is much easier for a factory to dump toxic WASTEWATER into the local river instead of spending money on purification procedures. However, the community suffers if the river becomes too polluted to be used for drinking or agriculture. The community may then consider adopting a law against polluting the river. Then the factory owners may threaten to move to a place that has fewer environmental laws. Such a move would result in the loss of jobs in the community.

A new environmental law always causes economic changes. These changes are often beneficial. Think of all the new products, such as degradable PLASTICS, that have been introduced in recent years. In fact, the environmental industry in the United States provides thousands of jobs and does business worth more than $100 billion a year. The industry includes companies that build water treatment plants, and pollution control devices for smokestacks and motor vehicles, as well as those that invent and manufacture environmentally friendly products. Manufacturers often have to change production methods to comply with new laws. Such changes may prove to be money savers. For instance, printers once used highly toxic inks and photographic products, and they had to pay for the removal and disposal of the products. Now, they save money by using BIODEGRADABLE products that produce no TOXIC WASTES. [*See also* ENVIRONMENTAL PROTECTION AGENCY (EPA).]

◆ Environmental laws exist that mandate cleaning up of existing pollution, as well as prevention of future pollution.

Law of the Sea

❚ A treaty adopted in the United Nations in 1982 that outlines practices for using the world's OCEANS. Earth's land has been divided up into countries and governed by laws for many centuries. In contrast, the two-thirds of Earth's surface covered by ocean has not been governed strictly. Over the centuries, people have fought over access to the sea and its resources. However, the ocean has been generally explored and exploited as if it were common property.

Knowledge of the oceans has changed greatly in the past century. The oceans are now mapped in detail. Even the most remote areas of the ocean bottom can be reached by people or their machines. Oil and MINERALS are extracted from the ocean floor.

Commercial fishing has become so widespread that entire FISH populations have been hunted nearly to the point of EXTINCTION. People are taking more and more food and resources from the ocean. At the same time, oceans everywhere are now affected by POLLUTION. Because of the ways people use and change the oceans, nations need detailed agreements on how oceans will be managed and protected.

THE LANGUAGE OF THE ENVIRONMENT

continental shelves parts of continents that extend for some distance under water beyond the continents' seacoasts. On a continental shelf, which may be a few miles or a few hundred miles across, the water is fairly shallow, and there are often large populations of fish and other marine animals. At the edge of the shelf, there is usually a steep drop to deep ocean.

landlocked enclosed by land.

◆ The Law of the Sea Convention is one of many attempts to regulate international use and abuse of the oceans. If ratified, it would affect such matters as marine pollution and the mining of seabeds.

THE CONVENTION ON THE LAW OF THE SEA

The United Nations adopted a treaty in 1982 called the Convention on the Law of the Sea (or the Law of the Sea Convention [LOSC]). The convention treaty includes the following goals and practices:

• It defines areas under control of one nation's laws, such as **continental shelves** and islands.

• It states that coastal nations have a right to find, use, and manage NATURAL RESOURCES located off their shores.

• It states that coastal nations have a duty to protect and manage the marine ENVIRONMENT, including living things.

• It states that **landlocked** nations have rights of access to the seas.

• It sets guidelines for international uses of the seas.

• It states the need for international laws on MARINE POLLUTION and for ways to enforce those laws.

• It states that ocean floor resources beyond national boundaries are a "common heritage" of humankind and should be developed according to rules set by the convention.

Many of the activities related to the Law of the Sea Convention are overseen by the United Nations Office for Ocean Affairs and the Law of the Sea. This office also helps form other policies that have to do with ocean affairs.

CURRENT STATUS OF THE LOSC

The Law of the Sea Convention is part of a long-term effort to manage the oceans. The Convention itself was the result of many years of work; the first United Nations Law of the Sea Conference had been held 24 years earlier, in 1958. Many of the provisions of the Law of the Sea Convention were not intended to go into effect until the 1990s.

Although this treaty has been adopted by the United Nations and has the support of many environmentalists, it may not go into full effect as international law. First, 60 nations must sign and ratify the agreement, and that may not happen. For example, the United States has not signed or ratified the treaty because it wants to keep certain rights to mine the seabed for manganese and other resources. Thus, the LOSC may have to be amended before it will be ratified by all 60 nations.

Despite its drawbacks, the Convention has served as an example for the shaping of other international agreements. Agreements about OCEAN DUMPING, exclusive economic zones that give countries special rights in certain areas, and guidelines for protection of the ocean environment have all resulted from work of the LOSC. [*See also* INTERNATIONAL CONVENTION FOR THE REGULATION OF WHALING (ICRW); SEABED DISPOSAL; and UNITED NATIONS ENVIRONMENTAL PROGRAMME (UNEP).]

Leaching

▶ The process of washing out and removing soluble materials, such as MINERALS, from the SOIL by moving water. Leaching, along with the amount of rainfall, determines the quality of the soil in many areas.

Without leaching, plant nutrients and minerals build up in soil. Thus, dry areas such as DESERTS tend to have fertile soils with high mineral contents. At the same time, the lack of rainfall in such areas restricts or limits the growth of PLANTS. Areas that receive moderate rainfall, such as GRASSLANDS and DECIDUOUS FORESTS, usually maintain a balance of soil minerals and water that fosters plant growth.

Leaching can sometimes result in poor soil quality. In areas receiving heavy rainfall, such as tropical RAIN FORESTS, leaching quickly washes substances out of the soil. The large numbers of plants living in these areas also tend to remove important nutrients from the soil as soon as they are formed. Thus, while tropical rain forests support a vast amount of vegetation, their soils tend to be nutrient poor. Leaching also causes CONIFEROUS FORESTS to have poor soils. These areas receive only moderate rainfalls, but the decaying needles on the forest floor make the rainwater and soil acidic. This acidic water also removes nutrients, such as iron, from the soil, making it infertile.

Leaching affects more than plants. Mining wastes are often placed in heaps called *spoil piles.* Spoil piles often contain low-grade ores and mineral wastes called TAILINGS. Rainwater washing through tailings can carry heavy metals and other harmful substances contained in the tailings into the groundwater supply or to nearby ponds or streams. Such leaching pollutes both land and water. Similar leaching activity also affects soil and water located near LANDFILLS.

Lead

▶ A soft blue-gray element (chemical symbol Pb) that is one of the world's oldest known metals and is important for use in the nuclear energy, chemical, and PETROLEUM industries. Pure lead is too soft for many uses and is frequently alloyed, or combined, with other metals, such as tin, to increase its strength. The exceptional flexibility of lead allows it to be forged into thin sheets or stretched without breaking. Among lead's other useful properties is its high resistance to corrosion by water, acids, and other chemicals and the protection it affords from sources of RADIOACTIVITY.

USES OF LEAD

Lead is widely used in the making of lead-acid batteries that store and provide power for electrical systems in vehicles. Lead compounds in paints used on bridges and other steel structures prevent **corrosion**. This corrosion-resistant capacity of lead also explains why lead is used to make pipes and tanks that are used to store and ship chemicals. Solder is a lead-tin **alloy** that is used to connect metal surfaces in cars and electronic devices.

At one time, lead was used in tetraethyl lead, a gasoline additive that increased auto-engine performance. However, the high **toxicity** of lead and its emission from exhausts has led to the development of unleaded fuels for use in vehicles. AUTOMOBILES that use leaded fuel emit chemicals that increase AIR POLLUTION. Because lead provides protection from RADIATION, it is used to line the walls of x-ray rooms and other places where radiation is present. Also, RADIOACTIVE WASTE is removed and stored in lead containers.

OBTAINING LEAD

Lead is extracted from galena, a gray metallic ore that contains lead and sulfur in pure form. Usually,

however, silver, COPPER, gold, and zinc are also present in galena. Through a series of processes involving chemicals and heat, the copper and other MINERALS are dissolved to recover the lead that remains.

The amount of lead that is used worldwide annually, 6 million tons (5.4 million metric tons), exceeds the production of 3.5 million tons (3 million metric tons). The additional lead is obtained through recycling, mostly of car batteries and lead pipes.

HAZARDOUS HELPER

Lead is classified as a heavy metal. The good news about lead is that it is very useful. The bad news is that it is extremely toxic and very harmful in the human body. Lead poisoning, a disease caused by extremely high levels of lead in the body, can cause **brain dysfunction** that leads to **coma**, **convulsions**, or death. Large amounts of lead interfere with the production of red blood cells and may damage not only the brain but also the liver, kidneys, and other organs. Severe cases of lead poisoning may be fatal, but such cases are rare. However, danger from lead poisoning increases if lead accumulates in the body over a long period of time. Such overaccumulation of lead is especially a problem for people who work in chemical plants, factories, or refineries in which lead particles become airborne in dust and fumes.

Lead may be taken into the body by inhalation, absorption through the skin, or swallowing. Lead may be swallowed by eating

◆ Containers made of lead are used to store waste material from nuclear power plants.

◆ Lead combines with sulfur to form lead sulfide, also called *galena*.

FISH taken from waters polluted by TOXIC WASTE from factories or, as can be the case with small children living in old buildings, when eating paint chips with high lead content.

Because of the dangers lead can cause to people, the government now regulates the lead levels allowed in products. Such legislation is responsible for the change from leaded to unleaded gasoline and the reduction in the use of lead pipes to carry water that may be used for drinking. In addition, the use of lead in paints has been largely eliminated. Even with such legislation, many environmentalists believe better enforcement of regulations is needed to stop toxic waste containing lead and other heavy metals from contaminating air, water, SOIL and WILDLIFE.

◆ Soybeans are legumes. Note the swellings or nodules on the roots, where bacteria fix nitrogen from the air.

Legumes

▶ PLANTS, such as peas, beans, peanuts, and soy beans, that form pods. Sometimes called the pea family, legumes include small plants, as well as shrubs and trees such as the mimosa. The seeds of small plants—bean, pea, soybean, and lentil—are staple food plants. Other legumes provide medicines, chemicals, cattle food, and timber.

Legumes have a butterflylike flower that forms a pod containing seeds. The leaves are usually arranged opposite one another along the stem. Their roots, which can reach to great depths, have small nodules or swellings. Within these nodules live *Rhizobium* BACTERIA, which are the nitrogen fixing organisms. They take nitrogen from the air and convert it into nitrogen compounds. The plant then uses the nitrogen compounds for growth, and the bacteria uses the sugar made by the plant.

When the legume dies and decays, the nitrogen compounds are added to the SOIL, thus enriching it. Farmers, aware of the soil-enriching quality of legumes, will often plant them during one season just to renew the nitrogen in the soil. [*See also* NITROGEN FIXING.]

Leopold, Aldo (1886–1948)

▶ **A**merican wildlife biologist and conservationist who began the practice of applying ecological standards to WILDLIFE MANAGEMENT. Born in Burlington, Iowa, Leopold attended Yale University. He graduated from Yale in 1908 but remained there to earn a master of forestry degree the following year. He immediately took a job with the U.S. FOREST SERVICE, working for the service from 1909 to 1927. In 1933, Leopold accepted a professorship at the University of Wisconsin and remained on the faculty there until his death in 1948. His biography, *Aldo Leopold: His Life and Work,* published in 1987, tells about Leopold's work in CONSERVATION.

A confirmed nature lover, Leopold believed that everyone should have the right to enjoy WILDERNESS areas, but only if the characteristics of such places could be preserved. He believed that if someone were to go to a wilderness area expecting clean air, clean water, abundant WILDLIFE, and peace and quiet— normal characteristics of such a place— that is what he or she should find.

Leopold wrote many articles and books emphasizing the need to conserve our NATURAL RESOURCES. *Game Management* (1933), a textbook that Leopold wrote on the subject of conservation, is considered by many to be timeless. In *A Sand County Almanac*, Leopold described the need for a new approach to humanity's ethical treatment of the ENVIRONMENT, an approach called the "land ethic." [*See also* ENVIRONMENTAL ETHICS.]

Lichen

▶ **A**n organism formed by a fungus growing together with an alga. The fungus forms a crusty or leaflike structure enclosing the alga.

Both FUNGI and ALGAE are simple PLANTS that lack true stems, roots, or leaves. Algae, unlike fungi, contain chlorophyll and perform PHO-TOSYNTHESIS, or use the sun to make food. Fungi are a group of organisms that live by absorbing water and nutrients from the ENVIRONMENT and from other organisms, either live or dead.

There are more than 20,000 kinds of lichens. They vary in appearance from the thin, crusty growths on a rock to stalks topped with domes to mosslike growths.

The fungi and the algae live together cooperatively, in a symbiotic relationship. Like other green plants, the algae make food through photosynthesis and pass the food on to the fungi. The fungi absorb water from the ENVIRONMENT. This process allows lichens to grow.

LICHENS AND THE ENVIRONMENT

Lichens play a valuable role in the formation of SOIL because they make acids that crumble rock. Some lichens expand when wet and get smaller when dry, fracturing rock in the process. The rock dust, or *regolith*, is the basic ingredient of soil. The continuous action of lichen on rock produces NICHES where other plants, such as moss, can take hold.

Because different lichens have very specific environmental needs, they are very good environmental indicator plants. Determining the kind of lichen tells a lot about the environment in which they are growing. Many lichens cannot live where there is sulfur gas or OZONE produced from AUTOMOBILE exhaust and thus are indicators of POLLUTION. The presence of lichens can indicate a pollution-free area. [*See also* SYMBIOSIS.]

◆ Note the two different kinds of lichen. On the left, fungus forms the outer part of the lichen; inside is an alga. Below, lichens growing on rock produce chemicals that crumble the rock.

Life-Cycle Assessment

▮▶ The scientific study used to determine the potential environmental effects of a new product or chemical. The life-cycle assessment, performed by the U.S. ENVIRONMENTAL PROTECTION AGENCY (EPA), is a critical evaluation of a product at each stage of its life cycle. For example, consider a new floor-cleaning chemical. A life-cycle assessment of such a product might include:

1. evaluating the environmental effects of the MINING and processing of the MINERALS and raw materials contained in the product;

2. assessing the environmental impacts of transporting and distributing the product; and

3. determining the potential environmental effects of using, reusing, and recycling the product and its packaging.

A typical life-cycle assessment has three parts: an inventory, an

◆ A life-cycle assessment determines the environmental impacts of a product from its production through its disposal.

ENVIRONMENTAL IMPACT STATEMENT, and recommendations. The life-cycle inventory lists the amounts and types of polluting substances a product contains at every stage of its life cycle. For example, for a new type of AUTOMOBILE, a life-cycle inventory will list the types of pollutants of the automobile, as well as the exact amounts of all gases released by its exhaust system.

The environmental impact analysis may be the most important part of a product's life-cycle assessment. In an environmental impact analysis, the short-term and long-term effects of each substance listed in the inventory are determined. This analysis may be long and complex because scientists must consider many different environmental factors. For example, scientists must consider all the ways people might use a product. They must also consider the ways people dispose of things.

The third part of the life-cycle assessment is a list of recommendations for the new product. The recommendations are suggested ways to improve the product and limit potential environmental dangers. For example, a recommendation may suggest changing the packaging materials for a product to reduce wastes.

The EPA is currently seeking ways to improve the procedure for evaluating new products. Meanwhile, the importance of life-cycle assessments will continue to grow, as industries try to improve product quality and at the same time comply with environmental laws. [*See also* AIR POLLUTION; BIOACCUMULATION; BIODEGRADABLE; PRIMARY POLLUTION; and WATER POLLUTION.]

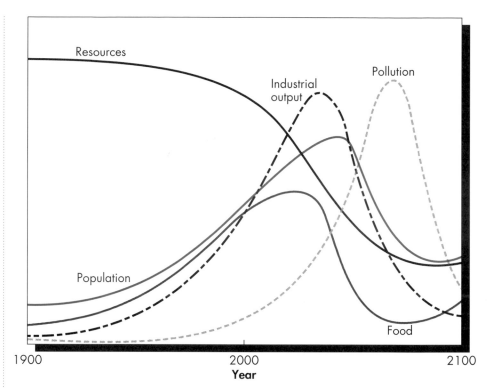

◆ A study by a group of industrialists concluded that industrial output pollution, food supply, and resources will peak in the twenty-first century, causing the collapse of industrial economies and rapid decline in population.

Limits to Growth

❙❙Conditions that affect the growth within a population of PLANTS or animals, including humans. A population cannot have a continuous unlimited growth because resources in its ENVIRONMENT are limited. A population grows only until it reaches its CARRYING CAPACITY, the number of members the environment can sustain. Any future growth of the population is delayed or stopped by outside forces called *limiting factors.*

WHAT CAN LIMIT GROWTH...AND HOW?

Density-dependent factors are outside forces that affect populations as they grow denser and more crowded. For example, disease, food shortages, and PREDATORS limit POPULATION GROWTH by increasing the number of deaths, reducing reproduction, and thus reducing the number of births.

Other outside forces, called *density-independent factors,* influence a population no matter its size. For example, NATURAL DISASTERS, HABITAT LOSS, and WEATHER conditions also increase deaths and limit reproduction, but they do so in both dense and sparse populations. For example, a flood might kill most of a cattle herd, but the size of the individual herd does not matter; the flood would affect any herd the same way.

Density-independent factors cause some populations to exist in constant "boom-and-bust" growth patterns—rapid population growth followed by a rapid decline in population. This yo-yo existence occurs naturally in some insect populations that constantly grow until they pass their carrying capacities. They eat all the food resources available, then the majority of the population starves to death. The surviving insects begin rebuilding toward another population boom, followed by another bust.

HUMAN ACTIVITY CAN BOTH LIMIT AND INCREASE GROWTH

Like other animal SPECIES, the human population is controlled by limiting factors such as disease, natural disasters, and food shortages. But humans have both limited and increased growth of their population through human activities. Activities such as wars and birth control have limited growth, while new medical procedures and health care practices, technology, and agricultural advances have increased growth. However, some human growth-increasing activities can create environmental problems that could limit or even end human population growth. For example, new medicines and better health care may allow people to live longer, and advances in technology may allow farmers to grow more food. But human populations can face dangers from density-dependent growth limitations if our population becomes dense enough to allow the spread of disease, cause food shortages on a regional or global scale, or change the global climate. [*See also* ECOSYSTEM; FAMILY PLANNING; FAMINE; and OVERPOPULATION.]

Lithosphere

The rocky, outermost layer of Earth, including the crust and the upper portion of the mantle. On average, the lithosphere is about 62 miles (100 kilometers) thick. The crust is the thin outer shell, ranging in thickness from about 3 to 5 miles (4.8 to 8 kilometers) under the OCEANS to about 12 to 56 miles (19 to 90 kilometers) on the continents. The mantle, which is Earth's thickest layer, lies just below the crust. It is composed mostly of silicon, OXYGEN, magnesium, and iron, and it flows like a liquid.

The structure of the lithosphere can best be understood by studying PLATE TECTONICS. Earth's crust is divided into about 15 sections, or "plates," that fit together like pieces in a jigsaw puzzle. These plates float and move in the fluid inner layers of Earth much as cereal floats in milk.

Processes in the lithosphere produce a variety of geological events on our planet, including earthquakes, mountain building, and volcanic eruptions. When the plates move against one another, tremendous energy is released at the plate boundaries. For example, when the Pacific plate moves past the North American plate, earthquakes may occur in California and volcanoes may erupt in Alaska. [*See also* CONTINENTAL DRIFT and VOLCANISM.]

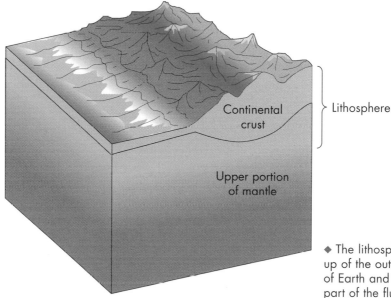

Continental crust

Lithosphere

Upper portion of mantle

◆ The lithosphere is made up of the outer, rigid crust of Earth and the upper part of the fluid mantle.

Livestock

▮Domesticated animals, such as hogs, cattle, sheep, poultry, and horses, raised for food and other valuable products. The raising and breeding of livestock is a science called *animal husbandry.*

Other livestock animals include rabbits, goats, donkeys, mules, llamas, reindeer, water buffalos, and yaks. Livestock provide us with such food products as eggs, meat, milk, butter, and cheese. The animals' hair, hides, and wool are used to make clothing, shoes, and brushes. Their feathers fill pillows, comforters, and insulated gloves and coats. Livestock hooves and horns are used to make combs, buttons, knives, and glue. Even their glands and organs are used to make drugs like insulin, and their fat is used for making soap and shortening. Livestock also provide vast amounts of body waste that can be used to fertilize SOIL and increase the growth of PLANTS.

THE HISTORY OF ANIMAL HUSBANDRY

Long ago, most humans were nomadic hunters and gatherers—people who wandered from place to place in search of food. Then, about 8,000 years ago, humans first domesticated, or tamed, the goat, which was quickly followed by the taming and breeding of other livestock. Raising their own animals instead of HUNTING for them allowed people to settle down in one place. People who settled down also began to farm.

◆ Raising cattle in feedlots uses less grazing area and provides farmers with large quantities of manure to fertilize crops naturally instead of chemically.

◆ Overgrazing leaves the soil bare and subject to wind and water erosion.

LIVESTOCK AND THE ENVIRONMENT

In some places, cattle, sheep, and other livestock roam free in pasturelands, GRAZING on the grass. OVERGRAZING can occur if too many animals feed in one pasture or if the animals are not rotated between grazing areas to allow for new plant growth. Overgrazing leaves the soil bare and subject to wind and water EROSION. Through erosion, nutrients for new growth are carried off with the TOPSOIL.

Animal wastes from hogs, chickens, and other livestock can pollute water supplies. After a storm, rainwater RUNOFF containing animal wastes flows into ponds and streams and, eventually, into larger waterways. FISH in the ponds and streams must then compete with the BACTERIA from organic wastes for life-giving OXYGEN. This conflict affects the distribution of life-forms in the water.

Today many farmers raise cattle in fenced-in areas known as *feedlots*. Here animals are fed special foods to increase their growth and to get them to market faster. Sometimes additives or drugs, like antibiotics and synthetic hormones, are added to livestock feed to increase their rate of growth and to prevent disease. In the United States, the government regulates the use of drugs that might harm either the animals or humans.

Feedlots require less grazing land. They also consolidate large amounts of animal waste that farmers can use in their fields to fertilize crops without using chemicals. If livestock consume INSECTICIDES and other chemicals sprayed on plants, these substances are often stored in the animal fat and thus passed along the FOOD CHAIN to humans who eat the livestock. Unfortunately, sometimes excess animal waste may run off feedlots into streams adding to POLLUTION problems. [*See also* BIOLOGICAL COMMUNITY; COMPETITION; DEFORESTATION; GRASSLANDS; ORGANIC FARMING; PUBLIC LAND; RANGELAND; and WATER POLLUTION.]

Loam

▶ A SOIL mixture with a medium texture that is well suited to plant growth. Loam consists of particles of sand, silt, and clay.

Clay, the finest soil particle, is the smallest component of loam. The sand and silt are therefore the primary materials that give loam its texture. In addition to sand, silt, and clay, loam contains a layer of HUMUS—partially decayed organic material. Humus provides nutrients needed for plant growth. Because of its texture, loam is able to hold water. This quality of loam, along with the presence of humus, makes loam an excellent soil for the growth of most PLANTS.

◆ Loam and humus provide an excellent soil for plant growth.

Love Canal

▶ The location of a chemical disposal landfill site in Niagara Falls, New York. Between the early 1940s and 1953, over 40,000 tons (36,000 metric tons) of chemical waste, much of it carcinogenic and toxic, were dumped at the site. A school and a housing development were later built on the LANDFILL.

In 1977, air, water, and SOIL around the Love Canal site were found to be heavily contaminated with 82 different chemicals. Exposure to toxic and carcinogenic chemicals was blamed for the high incidence of health problems and genetic damage among people living on the site. During the next

◆ In the 1990s, people started to move back to parts of Love Canal that were considered safe.

three years, hundreds of families were evacuated at a cost of $27 million, and the site was fenced off and declared a federal disaster area. This and other incidents have resulted in the current strategies to dispose of HAZARDOUS WASTES safely, to recycle them, and to reduce their production.

After testing the outskirts of the site in the 1990s, part of the land was deemed habitable. By the mid-1990s, people were moving into this once condemned area. [*See also* CARCINOGEN; DIOXIN; HAZARDOUS WASTE MANAGEMENT; HAZARDOUS WASTE, STORAGE AND TRANSPORTATION OF; INDUSTRIAL WASTE TREATMENT; LANDFILL; LAW, ENVIRONMENTAL; and TOXIC WASTE.]

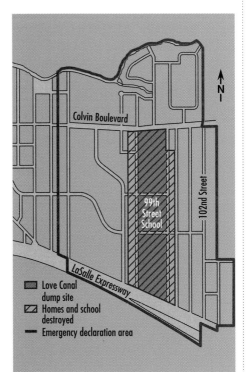

◆ As indicated in the diagram above, the area affected by the Love Canal dump site was large, including many homes and a school.

Lovelock, James (1919–)

▮ The British chemist best known for the development of the GAIA HYPOTHESIS. An author of over 200 scientific papers, Lovelock is also known for the development of the electron capture gas chromatographic detector. His invention measures the level of CFCs and other substances in the ATMOSPHERE.

EDUCATION

Lovelock was born in England in 1919. At Manchester University, Lovelock earned a degree in chemistry, and he went on to earn a Ph.D. in medicine from the London School of Hygiene and Tropical Medicine. Later, he spent some time studying medicine at Harvard Medical School and Yale University. His most significant work as a scientist occurred while working at NASA's Jet Propulsion Laboratory.

GAIA HYPOTHESIS

Lovelock's Gaia Hypothesis was developed while he was at NASA, in collaboration with the noted biologist Lynn Margulis. The Gaia Hypothesis is a model for how Earth and its organisms function together as a unit. Gaia was the ancient Greek god of Earth. According to this hypothesis, the Earth and all of its living things work together as one system. Under this hypothesis, changes to any one part of the system can affect the entire system, much like an injury or illness to one part of your body can make your entire body feel sick.

USING THE HYPOTHESIS

Lovelock's hypothesis has received some criticism. However, the Gaia Hypothesis has had an impact on the scientific community. Many scientists feel that the Gaia Hypothesis is a useful model for testing ideas. Some studies have even given results predicted by the Gaia Hypothesis.

Today, Lovelock continues his research on the Gaia Hypothesis. He has been honored by the scientific community worldwide and has received numerous awards, including the prestigious Fellow of the Royal Society award. [*See also* OZONE HOLE and OZONE LAYER.]